Global Monitoring Report 2012

Food Prices, Nutrition, and the Millennium Development Goals

Food Prices, Nutrition, and the Millennium Development Goals

The painting on the cover is by Sue Hoppe, an artist based in South Africa. Titled "Resolution," the painting explores the idea that people who seem irreversibly divided and with little in common can unite if they focus on what they have in common instead of what divides them. Hoppe's work examines war, conflict, and the plight of children and women in Africa, but is also inspired by nature and architecture. To learn more about Sue Hoppe and her work, visit www.southafricanartists.com/home/SueHoppe.

Cover design by Debra Naylor of Naylor Design

Photo credits: page xvi: Masuru Goto/World Bank; clockwise for pages 10–11, beginning at top: Liang Qiang / World Bank, Curt Carnemark / World Bank, Curt Carnemark / World Bank, and Steve Harris / World Bank; page 15: Curt Carnemark / World Bank; page 16: Curt Carnemark / World Bank; page 19: Curt Carnemark / World Bank; page 21: John Isaac / World Bank; page 23: John Isaac / World Bank; page 25: Curt Carnemark / World Bank; page 27: Curt Carnemark / World Bank; page 28: Michael Foley; page 62: Arne Hoel/World Bank; page 94: Shehzad Noorani/World Bank; page 116: Alex Baluyut/World Bank; page 136: Michael Foley.

Contents

BOXES

MAPS

TABLES

Foreword

Every year, the *Global Monitoring Report* (GMR) gauges progress across the Millennium Development Goals (MDGs), so we can better understand whether we are delivering on basic global needs. These needs include affordable, nutritious food; access to health services and education; and the ability to tap natural resources sustainably—whether clean water, land for urban expansion, or renewable energy sources. We assess how well the world is doing by looking at income poverty, schooling levels, the health of mothers and children, and inroads in treating HIV/AIDS, malaria, and tuberculosis, as well as assessing how the international development community delivers aid. We also try to measure levels of malnutrition and hunger in the world. Food prices can affect all these indicators.

For these reasons, the *Global Monitoring Report 2012* takes the theme of *"Food Prices, Nutrition, and the Millennium Development Goals."* This year's edition highlights the need to help developing countries deal with the harmful effects of higher and more volatile food prices.

In 2007–08 and again in 2011, soaring food prices held back millions of households from escaping poverty. Poor people in cities remain especially vulnerable to higher food prices, as do households headed by women. Higher food prices also affect the quantity and quality of nutrition—a critical factor for children in the first two years of life, when even a temporary reduction in nutritional intake can affect long-term development. This loss of nutrition can, in turn, set back a whole generation.

The GMR details some of the solutions for making countries and communities more resilient in the face of food price spikes. Strategies include using agricultural policies to encourage farmers to boost production; using social safety nets to improve resilience; strengthening nutritional policies to manage the implications of early childhood development; and designing trade policies to improve access to food markets, reduce food price volatility, and make productivity gains.

The implications of high and more volatile food prices vary widely at the regional and country levels. Large net importers of food—such as those in the Middle East, North Africa, and West Africa—face higher import bills, reduced fiscal space, and greater transmission of world prices to local prices for imported rice and wheat. Higher prices hurt consumers, who need to spend a greater share of their income on food, as is the case in much of Africa and Asia. Larger net exporting countries, such as those in Latin America, Eastern Europe, and Central Asia, stand to benefit. But they may also face internal pressure to help households that need to spend a large share of their budgets on food. The sequencing and prioritization of policy initiatives depends critically on a country or region's initial situation.

Going forward, all of us—including traditional donors, new donors, philanthropists, and NGOs—must do better in fighting hunger, particularly by making more resources available for basic nutrition. For a start, this means including nutrition interventions in projects and programs when and wherever possible. At the same time, we need to design more effective policies, strengthen accountability, and ensure that recipients can absorb vital assistance.

The GMR's assessment of progress on the MDGs offers grounds for optimism. Global targets for overcoming extreme poverty and access to safe drinking water have been reached well ahead of schedule. Goals related to primary school completion rate and gender equality in primary and secondary education also appear within reach. Other goals, however, require a real push, particularly regarding child and maternal mortality, and access to improved sanitation facilities. MDG gaps are starker when the focus is on individual countries and achievements per region, where disparities persist.

Macroeconomic performance will play a critical role in meeting the MDGs. Progress that was made possible by the relatively strong economic growth of developing countries prior to the global financial crisis has been set back. The recent weakening of the global economic environment has implications for overcoming poverty in emerging and developing economies, and it is important that the advanced economies undertake the necessary macroeconomic policies to bring about strong and stable global growth.

A key concern lies with the low-income countries, where macroeconomic policy buffers—such as fiscal, debt, and current account positions—have not yet been rebuilt to levels before the crisis. If they have to confront another sharp global slowdown or another surge in food or fuel prices, these countries would start from a weaker position.

We have made important progress in pushing forward toward meeting the MDGs—but the year 2015 is just around the corner. We have three years to ensure that billions more people will have the opportunity to benefit from the global economy. The need for cooperation on focused steps to achieve these goals has never been greater.

Robert B. Zoellick
President
The World Bank Group

Christine Lagarde
Managing Director
International Monetary Fund

Acknowledgments

This report has been prepared jointly by the staff of the World Bank and the International Monetary Fund. In preparing the report, staff have collaborated closely with partner institutions—the African Development Bank, the Asian Development Bank, the European Bank for Reconstruction and Development, the Inter-American Development Bank, the Organisation for Economic Co-operation and Development, the Food and Agriculture Organization, the European Union, the UK Department for International Development, and various NGOs, such as Oxfam International and Save the Children. The cooperation and support of the staff of these institutions is gratefully acknowledged.

Jos Verbeek was the lead author and manager of the report. Lynge Nielsen led the team from the IMF. The principal authors and contributors to the various parts of the report include Mohini Datt, Annette I. De Kleine Feige, Ian Gillson, Rasmus Heltberg, Maros Ivanic, Bénédicte de la Briere, Hans Lofgren, Maryla Maliszewska, William J. Martin, Jose Alejandro Quijada, Eric V. Swanson, and Sergiy Zorya (World Bank), and Sibabrata Das, Stefania Fabrizio, Yasemin Bal Gunduz, Svitlana Maslova, and John Simon (IMF). Sachin Shahria assisted with the overall preparation and coordination of the report. The work was carried out under the general guidance of Justin Yifu Lin and Hans Timmer at the World Bank. Supervision at the IMF was provided by Hugh Bredenkamp and Brad McDonald.

A number of other staff and consultants made valuable contributions, including the following from the World Bank: Abebe Adugna, Harold Alderman, Lystra N. Antoine, Jean Francois Arvis, John Baffes, Saswati Bora, Andrew Burns, Grant Cameron, Gero Carletto, Iride Ceccacci, Shaohua Chen, Loriza Dagdag, Christopher Delgado, Asli Demirgüç-Kunt, Leslie Elder, Neil Fantom, Ariel Fiszbein, Delfin Sia Go, Anna Herforth, Masako Hiraga, Hans Hoogeveen, Alma Kanani, Norman Loayza, Alessandra Marini, Dominique van der Mensbrugghe, Menno Mulder-Sibanda, Israel Osorio-Rodarte, Martin Ravallion, Anna Reva, Bruce Ross-Larson, Julie Ruel Bergeron, Cristina Savescu, William Shaw, Meera Shekar, Yurie Tanimichi Hoberg, Robert Townsend, Jonathan Wadsworth, and Ruslan Yemtsov.

Contributors from other institutions included: Duncan Green and Richard King (Oxfam); Naomi Hossain (Institute of Development Studies); Kate Dooley and Daphne Jayasinghe (Save the Children); Fredrik Ericsson, Kimberly Smith, and Suzanne Steensen (OECD); Indu Bhushan (Asian Development Bank); Amy M. Lewis (Inter-American Development Bank); Jennifer Keegan-Buckley and Jean-Pierre Halkin (European Union); Chris Penrose-Buckley (DFID); Murat Jadraliyev

(EBRD); Anita Taci (EBRD); and Patricia N. Laverley (Africa Development Bank).

Guidance received from the Executive Directors of the World Bank and the IMF and their staff during discussions of the draft report is gratefully acknowledged. The report also benefited from many useful comments and suggestions received from the Bank and Fund management and staff in the course of its preparation and review.

The World Bank's Office of the Publisher managed the editorial services, design, production, and printing of the report, with Aziz Gokdemir anchoring the process. Others assisting with the report's publication included Denise Bergeron, Susan Graham, Stephen McGroarty, and Santiago Pombo-Bejarano.

The report's dissemination and outreach was coordinated by Indira Chand and Merrell Tuck-Primdahl, working with Vamsee Kanchi, Malarvizhi Veerappan, and Roula Yazigi.

Abbreviations and Acronyms

ADB	Asian Development Bank		HIV	human immunodeficiency virus
AfDB	African Development Bank		IDB	Inter-American Development Bank
AIDS	acquired immune deficiency syndrome		IFC	International Finance Corporation
AMIS	Agricultural Market Information System		IFI	international financial institution
			IFPRI	International Food Policy Research Institute
BMI	body mass index			
BRICS	Brazil, Russia, India, China, and South Africa		LSMS	Living Standards Measurement Study
			MAMS	maquette for MDG simulations
CGIAR	Consultative Group on International Agricultural Research		MDB	multilateral development bank
			NTM	non-tariff measure
CPA	country programmable aid		OAP	Open Aid Partnership
DAC	Development Assistance Committee		ODA	official development assistance
DFID	Department for International Development (U.K.)		OECD	Organisation for Co-operation and Development
EBRD	European Bank for Reconstruction and Development		PFM	public finance management
			PPP	purchasing power parity
EU	European Union		PSE	producer support estimates
FAO	Food and Agriculture Organization (of the United Nations)		PSI	pre-shipment instructions
			REPO	repurchase option
FDI	foreign direct investment		RTA	regional trade agreement
G-8	Group of Eight		SPS	sanitary and phytosanitary
G-20	Group of 20		SUN	Scaling Up Nutrition
GATT	General Agreement on Tariffs and Trade		TBT	technical barriers to trade
			UNCTAD	United Nations Conference on Trade and Development
GFRP	Global Food Crisis Response Program			
GIDD	Global Income Distribution Dynamics		WFP	World Food Programme
GMR	*Global Monitoring Report*		WTO	World Trade Organization
GNI	gross national income			
GTAP	Global Trade Analysis Project			

All amounts are presented in U.S. dollars, unless otherwise indicated.

Overview

What has been the impact of yet another food price spike on the ability of developing countries to make progress toward the Millennium Development Goals (MDGs)? How many poor people were prevented from lifting themselves out of poverty? How many people, and how many children, saw their personal growth and development permanently harmed because their families could not afford to buy food? How did countries react to the last two food price spikes of 2007–08 and 2010–11, and how did their reaction affect their progress toward the MDGs? And what can countries do to respond to higher and more volatile food prices? The 2012 *Global Monitoring Report* (GMR) addresses these basic questions. It summarizes effects of food prices on several MDGs. It reviews policy responses—including domestic social safety nets, nutritional programs, agricultural policies, regional trade policies, and support by the international community. And it outlines future prospects.

The world has met two MDGs, while global progress varies across the other MDGs (figure 1). Preliminary survey-based estimates for MDG 1.a in 2010—based on a smaller sample than the global update in box 1—indicate that the $1.25 a day poverty rate (2005 purchasing power parity, or PPP)

had fallen below half its 1990 value in 2010. Also in 2010, the world met MDG 7.c—to halve the proportion of people with no safe drinking water—well ahead of the 2015 deadline. And global progress on various MDGs is on track or within 10 percent of the on-track trajectory. MDG 3.a (gender parity in school enrollment) is on track, and MDG 2.a (primary school completion) is close to being on track. But the MDGs closely linked to food and nutrition are lagging, particularly child mortality (MDG 4) and maternal mortality (MDG 5). The same is true for country progress: 105 countries of the 144 monitored are not expected to reach MDG 4, and 94 are off track on MDG 5.

Food prices spike once again

In 2011 international food prices spiked for the second time in three years, igniting concerns about a repeat of the 2008 food price crisis and its consequences for the poor. The World Bank Food Price Index rose 184 percent from January 2000 to June 2008 (figure 2). In February 2011 it again reached the 2008 peak, after a sharp decline in 2009, and stayed close to that peak through September. The international food price spike in 2007–08 is estimated to have kept or pushed 105

1

FIGURE 1   Global progress toward the MDGs varies

Developing countries, weighted by population

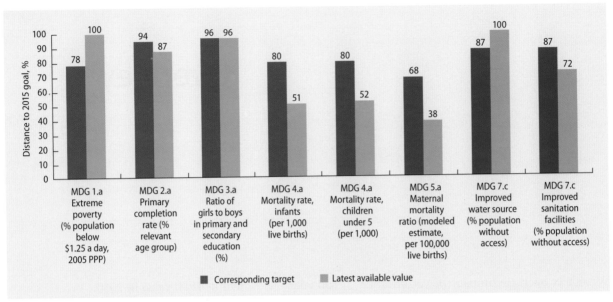

Source: World Bank staff calculations based on data from the World Development Indicators database.
Note: A value of 100 percent means that the respective MDG has been reached. "Corresponding target" indicates progress currently needed to reach the goal by 2015. "Latest available value" denotes current progress as illustrated by the most recent available data: extreme poverty, 2010; primary completion rate, total, 2009; ratio of girls to boys in primary and secondary education, 2009; mortality rate, infants, 2010; mortality rate, children under 5, 2010; maternal mortality ratio, 2008; improved water source, 2010; improved sanitation facilities, 2008. PPP stands for purchasing power parity.

BOX 1   The MDG target of halving extreme poverty—reached in 2010!

The World Bank has been regularly monitoring the progress of developing countries in reducing extreme poverty. Drawing on data and expertise from all regions, the Bank has updated the global and regional poverty numbers for 1981–2008 and prepared preliminary estimates (for a smaller sample) for 2010. The latest estimates draw on more than 850 household surveys for almost 130 developing countries, with 90 percent of the developing world population. Mostly produced by national statistical offices, the results for 2005 and 2008 are based on interviews with 1.23 million randomly sampled households.

An estimated 1.29 billion people in 2008 lived on less than $1.25 a day, equivalent to 22.4 percent of the developing world population (see the box table on the next page). Contrast that with 1.9 billion people in 1990, or 43.1 percent.

Preliminary survey-based estimates for 2010—based on a smaller sample than the global update—indicate that the $1.25 a day poverty rate had fallen

to less than half of its 1990 value by 2010. So the first MDG target of halving extreme poverty has been achieved well before the 2015 deadline. East Asia and Pacific, Middle East and North Africa, and Europe and Central Asia have attained MDG 1.a, while poverty in South Asia and Sub-Saharan Africa remains in double digits. Current estimates for 2015 show that poverty will further decline to 16.3 percent for the world as a whole.

Looking back to 1990, **East Asia and Pacific** was the region with the highest number of poor people in the world, with 926 million living below $1.25 a day. By 2008 that level had fallen to 284.4 million. In China alone, 510 million fewer people were living in poverty by the $1.25 standard. In 2008, 13 percent (173 million people) of China's population still lived below $1.25 a day. In **South Asia,** the $1.25 a day poverty rate fell from 54 percent to 36 percent between 1990 and 2008. The proportion of poor is lower now in South Asia than at any time since 1981.

BOX 1   The MDG target of halving extreme poverty—reached in 2010! (continued)

The number of poor had been generally rising in **Latin America and the Caribbean** until 2002. But the poverty count (and the percentage of poor) has fallen sharply since then. The rising incidence and number of poor in **Europe and Central Asia** has also been reversed since 2000. The **Middle East and North Africa** had 8.6 million people—or 2.7 percent of the population—living on less than $1.25 a day in 2008, down from 10.5 million in 2005 and 13 million in 1990. Less than half the population of **Sub-Saharan**

Africa was living below $1.25 a day in 2008. Forty-seven and a half percent lived below this poverty line in 2008, as compared with 56.5 percent in 1990, a 9 percentage point drop.

Good news, but a great many people remain poor and vulnerable in all regions. At the current rate of progress, around 1 billion people will still be living below $1.25 a day in 2015. Most of the 619 million poor lifted above the $1.25 a day standard during 1990–2008 are still poor by middle-income standards.

Estimates of poverty on a poverty line of $1.25, by region

Region	1990	2005	2008	2015
Share of population living on less than $1.25 a day (2005 PPP)				
East Asia and Pacific	56.2	16.8	14.3	7.7
of which, China	60.2	16.3	13.1	—
Europe and Central Asia	1.9	1.3	0.5	0.3
Latin America and the Caribbean	12.2	8.7	6.5	5.5
Middle East and North Africa	5.8	3.5	2.7	2.7
South Asia	53.8	39.4	36.0	23.9
Sub-Saharan Africa	56.5	52.3	47.5	41.2
Total	**43.1**	**25.0**	**22.4**	**16.3**
Total, excluding China	37.2	27.7	25.2	—
Millions of people below $1.25 a day (2005 PPP)				
East Asia and Pacific	926.4	332.1	284.4	159.3
of which, China	683.2	211.9	173.0	—
Europe and Central Asia	8.9	6.3	2.2	1.4
Latin America and the Caribbean	53.4	47.6	36.8	33.6
Middle East and North Africa	13.0	10.5	8.6	9.7
South Asia	617.3	598.3	570.9	418.7
Sub-Saharan Africa	289.7	394.9	386.0	397.2
Total	**1,908.6**	**1,389.6**	**1,289.0**	**1,019.9**
Total, excluding China	1,226.8	1,177.7	1,116.0	—

Source: World Bank staff calculations from PovcalNet database. For additional information and data, see http://iresearch.worldbank .org/PovcalNet/index.htm.
— = not available.

million people below the poverty line, and in the spike of 2010–11, 48.6 million people. Poverty typically rises initially with higher food prices, because the supply response to higher prices takes time to materialize and many poor (farm) households are net food buyers, so higher food prices lowers their real incomes.

The regional and national implications of high and volatile food prices vary widely. How vulnerable a country is to food price spikes depends on whether it is a net exporter or net importer. Large net importers of food, such as those in the Middle East and North Africa and in West Africa, face higher import bills, reduced fiscal space, and greater transmission of world prices to local prices for imported rice and wheat. Higher prices hurt consumers with high shares of household spending on food, as in much of Africa and Asia. Larger net-exporter countries, as in Latin America and in Eastern Europe

FIGURE 2   **Food prices spiked again for the second time in three years**

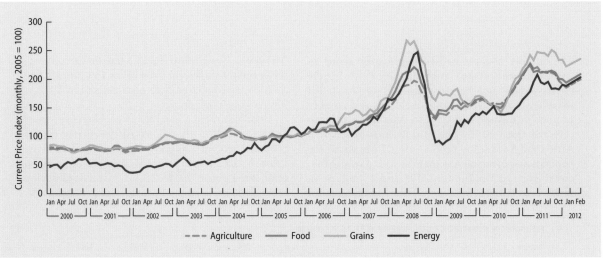

Source: World Bank.
Note: The World Bank Food Price Index includes wheat, maize, rice, barley, sugar, coconut oil, soybean oil, groundnut oil, palm oil, copra, soybeans, soybean meal, orange, banana, beef, and chicken. Unlike the well-known Food and Agriculture Organization food price index, it does not include other meat and dairy.

and Central Asia, stand to benefit. But they may also face internal pressure to mitigate the adverse effects if households spend large shares of their budgets on food.

A multisectoral approach is needed, tailored to each country's conditions, taking into account the social and political environment. This *Global Monitoring Report* advocates agricultural policy mainly to orchestrate a supply response, social safety nets to improve resilience, nutritional policy to manage the implications of early childhood development, and trade policy to improve access to food markets, reduce volatility, and induce productivity gains. But one size does not fit all. The priority and sequence of various policy initiatives depends on a country's or a region's initial conditions.

Combinations of policies in the four areas can provide positive synergies and spur improvements on the MDGs. Targeting the expansion and productivity of crops that add nutritional value is one example. At the same time, improvements in the value chain of food products through, say, investing in infrastructure and streamlining regulation related to trade can lead to faster pass-throughs of

international prices—and thus require an effort to improve resilience using social safety nets. Both require time to implement, so the sequencing of actions needs to avoid hardship for the poor and vulnerable.

How the poor cope

High and volatile food prices hurt food security. Large, sudden, and particularly unexpected food price increases make it difficult for households to adjust—eroding consumer purchasing power, reducing calorie intake and nutrition, and pushing more people into poverty and hunger. The poor bear a disproportionate burden in adjusting to high food prices. This is especially true for poor households in urban settings and those headed by women, who typically spend more than half their incomes on food and are more likely to curtail consumption in the face of higher prices.

The higher prices also have indirect effects. Poor people have experienced global shocks in recent years, from the spikes in fuel and food prices to the economic contraction starting in 2008 and the consequent reductions in remittances. And droughts have made things even worse in many countries and locales. Qualitative survey-based research shows that the responses of poor people to past global shocks produced severe indirect impacts.[1] Less nutritious diets caused malnourishment and made people more susceptible to failing health. The sudden influx of workers into the informal economy lowered earnings. The hardships even led to criminal activities, eroding trust and cohesion in communities.

Reducing the quality of food and the number of meals was one of the most common responses, often the first, in study sites in countries surveyed. In addition, reducing non-food consumption, working more hours, and diversifying income sources (say, by entering a new informal occupation) were common nearly everywhere. Migration, sometimes reverse migration to the home area, was also fairly common in response to the food price spikes. Asset sales were common, and loans from family, friends, and moneylenders were also important. An inability to service microfinance and moneylender debts was a major source of distress in some East and South Asian countries, where many people had to borrow at high interest rates to service these debts or live in fear of creditors taking possession of their property. Collecting food and fuel from common property was important only in some low-income countries.

Sales of productive assets and forgone education and health care will have long-lasting consequences and impede people's ability to recover. And coping with economic crises has eroded the savings and asset base of many households, leaving them with few resources to manage future shocks. Continuing high and volatile global food prices are thus a major concern.

Many parents sought to protect children's food consumption and schooling, with adult household members preserving the quantity and quality of food to ensure that children had proper diets. Yes, there were many instances of erratic attendance and school withdrawals because of the need for children to contribute to household income or because education costs had become prohibitive. But,

on the whole, the impacts on schooling were more muted than expected. The cost of education, the distance to school, and the availability of school feeding influenced whether children stayed in school.

What higher prices mean for poverty

The food price spikes in 2008 and 2011 have prevented millions of people from escaping poverty because the poor spend large shares of their incomes on food—and because many poor farmers are net buyers of food. The price spikes hit urban poor and female-headed households hardest. While food prices dropped sharply in 2009 with the financial crisis, they quickly rebounded and by early 2011 prices were almost back at 2008 levels. But high food prices may be less harmful for the poor in the longer run because farmers can increase their output and poor households may benefit from higher farm wages.

The impact of world price spikes also depends on how prices are transmitted locally. In Ethiopia, for example, about 75 percent of food consumption is locally produced staples (such as sorghum and teff), dampening the impact of rising prices of imported cereals. By contrast, people in Bangladesh, Cambodia, and Kenya—where rice, wheat, maize, and beans account for 40–64 percent of food expenditures—are more exposed to higher import prices. Changes in international prices have been trickling down to national prices to varying degrees, but the higher national prices have greatly influenced national policies.

In the longer run, farmers can be expected to respond to rising food prices in two ways: by raising their overall output and by switching to producing commodities whose prices have risen relative to others. For short-run price volatility, where producer outputs are likely fixed, the change in farmers' revenue from production is determined only by the change in output prices. But over time outputs can be expanded by using more labor and inputs, even if no additional land is available. Where the relative prices of different commodities change, switching between outputs is another way for farmers to increase returns. In general, it is easier for farmers to switch production to a more profitable crop than to increase aggregate agricultural output.

Policies that promote higher yields can limit the average rise in food prices over the long term as well as dampen food price volatility. Such policies include supporting research, extension, and water management; improving the efficiency of land markets and strengthening property rights; increasing farmers' access to efficient tools to manage risk; and increasing market integration, globally as well as regionally, through investments in infrastructure and facilitating the operations of supply chains. Policies to limit food price volatility include developing weather-tolerant varieties, improving the management of stocks, opening markets to trade, improving market transparency, and using market-based price-hedging mechanisms. And policies to reduce the impact of high and volatile food prices on the poor include strengthening a country's social safety nets to protect the poor and supporting smallholders in strengthening the supply response to higher food prices.

Balancing the rise in domestic prices (to benefit producers) with consumer protection is a major challenge. Because of fiscal constraints, many countries use trade measures to limit the transmission of higher world market prices to domestic markets. Scaling up safety nets to support vulnerable consumers without also insulating markets has been rare, hurting long-term food security. The most sustainable policies focus on encouraging climate-resilient production, strengthening domestic and regional markets, maintaining open trade, and boosting resources to social protection.

Higher undernourishment

Higher prices of food staples increase undernourishment, as poor consumers find themselves unable to purchase the minimum amount of calories, nutrients, and proteins needed for their daily activities. Higher food prices have two main effects on net buyers of food: an income effect through reductions in the purchasing power of poor households, and a substitution effect through shifts to less nutritious food. The poor often have no choice but to reduce their overall food consumption in response to higher prices, even from levels already too low. For households close to subsistence and already consuming the cheapest sources of calories (less nutritious food), the substitution possibilities are more limited, with the poorest suffering most. And intrahousehold discrimination against women and children disproportionately reduces their access to food.

Even temporarily high food prices can affect children's long-term development. Early life conditions (from conception to two years of age) provide the foundations for adult human capital. Vicious circles of malnutrition, poor health, and impaired cognitive development set children on lower, often irreversible, development paths. Child malnutrition accounts for more than a third of the under-five mortality—and malnutrition during pregnancy, for more than a fifth of maternal mortality. Other hard-to-reverse impacts include faltering growth (stunting, low height-for-age), and low school attainment. A malnourished child has on average a seven-month delay in starting school, a 0.7 grade loss in schooling, and potentially a 10–17 percent reduction in lifetime earnings—damaging future human capital and causing national GDP losses estimated at 2–3 percent. So, malnutrition is not just a result of poverty—it is also a cause. Malnourished young children are also more at risk for chronic disease such as diabetes, obesity, hypertension, and cardiovascular disease in adulthood.

To build household and individual resilience and mitigate long-term effects, interventions can work through multiple pathways, beyond trying to keep prices low. In the short run, the focus should be on maintaining household purchasing power through cash transfers, food and nutrient transfers, school feeding, and workfare-with-nutrition. In the longer term, the focus should broaden to strengthening the link between smallholder agriculture and nutrition, addressing seasonal deprivation, and promoting girls' education and women's income. Some specific interventions to target vulnerable children include hygiene, micronutrients, deworming, breastfeeding, feeding during illness, and preventive and therapeutic feeding.

Weaker global growth and high food prices may impede progress toward the MDGs

The global recovery shows signs of stalling amid deteriorating financial conditions. Global growth slowed to 3.8 percent in 2011 and is projected to decline further to 3.3 percent in 2012. The strongest slowdown is being felt in advanced economies, but the worsening external environment and some weakening in internal demand is expected to lead to lower growth in emerging and developing countries as well. This outlook is subject to downside risks, such as a much larger and more protracted bank deleveraging in the Euro Area or a hard landing by key emerging economies.

Strengthening the recovery will require sustained policy adjustment at a measured pace that depends heavily on a country's circumstances. There are risks in some places of inadequate medium-term fiscal adjustment, and in some of overly aggressive short-term fiscal adjustment. In the advanced economies, while fiscal policy consolidation proceeds, monetary policy should continue to support growth as long as unemployment remains high and inflation expectations are anchored. This policy stance should be accompanied by steady progress toward repairing and reforming financial systems and by steps to avoid excessively rapid bank deleveraging.

As food and fuel prices rose in 2010 and the first half of 2011, consumer prices rose in tandem in many countries. In emerging and developing countries the median inflation rate rose from 4 percent in 2009 to 6 percent in 2011, but experiences were mixed. In about a third of countries inflation abated

over this period, but in many countries it rose sharply. In Burundi inflation more than tripled from 4½ percent in 2009 to 15 percent in 2011 as the monetary authorities sought to contain the second-round effects of imported inflation. And in Bangladesh inflation doubled from 5½ percent to 11 percent.

A weaker-than-expected global economic environment would challenge emerging and developing countries as they progress toward the MDGs. Should downside risks such as a sharp global slowdown or another surge in food prices materialize, many low-income countries would have to confront the situation with weaker buffers than in 2009. In the event of another sharp downturn, the scope for fiscal stimulus would therefore be more limited, but those with sufficient fiscal room should aim to protect spending to soften the economic and social impact of a global downturn. A new global food price spike would present low-income countries with difficult trade-offs among price stability, external, and social objectives. A pragmatic response should include measures to protect the poor and vulnerable while largely accommodating the first-round impact on inflation. The fiscal policy response should be well targeted, ensure fiscal affordability, and avoid economic distortions. The appropriate monetary policy response would depend on the inflation outlook, the pass-through from food prices to other prices, and the availability of external buffers, such as reserves. Fragile states would require special attention including from the international community.

Using trade policy to overcome food insecurity

International cereal price spikes increased the food import bills of some low-income food-deficit countries, putting pressure on their balance of payments. The cereal import bill of low-income food-deficit countries was $31.8 billion in 2010–11 (29 percent more than in 2009–10), despite higher production in 2010 and the lower volume of cereal imports required. North Africa and the Pacific Islands suffered the largest negative impact, paying higher prices and importing more cereals to meet domestic demand. Although the estimated cereal import bill of the food-deficit countries is still below the record set during the 2008 food crisis, the increase in cereal costs, combined with price increases for other food and fertilizer imported by these countries, is cause for concern.

Higher food prices can upset the balance between needed spending to mitigate the immediate impact of the crisis and long-term development. Recurrent food crises may require additional social spending; to be cost-effective, such spending should emphasize targeted social safety nets rather than universal producer and consumer subsidies. Most developing countries preserved their core spending on health, education, and infrastructure during 2008–09, increasing their resilience to food and financial crises. In the period since, however, many countries have not adequately rebuilt their fiscal policy buffers and thus may find it more difficult to preserve core spending in the face of another major shock. To maintain this resilience in the composition of expenditures, much will depend on the cost and availability of resources going forward.

Trade is an excellent buffer for domestic fluctuations in food supply. There is no global food shortage: the problem is regional or local, one of moving food, often across borders, from surplus to deficit areas, coupled with affordability. The world output of a given food commodity is far less variable than the output in individual countries. So greater trade integration holds considerable potential for stabilizing food prices, boosting returns to farmers, and reducing the prices facing consumers.

Trade liberalization protects national food markets against domestic shocks by allowing more food to be imported in times of shortage and exported in times of plenty. But historically—and despite a host of regional trade agreements—most countries have taken the opposite approach. They restrict food imports and discourage exports in often-failed attempts to keep domestic markets

isolated from international shocks by ensuring self-sufficiency in food production.

Self-sufficiency should be weighed against the benefits of cheaper imports. A country that is a natural exporter should not encumber its comparative advantage with export bans. A country that tends to import food should allow its domestic market to remain linked to the world market. Encouraging more trade—not less—can thus promote food security, which requires a more open, rules-based multilateral trade regime best achieved by concluding the Doha Round of negotiations at the World Trade Organization.

Efforts to extend trade integration to developing countries should also focus on promoting more effective regional integration, including that for food products. Facilitating food trade is also important through increased Aid for Trade to promote frictionless borders and facilitate a supply response to rising prices, particularly in Sub-Saharan Africa.

Aid flows, composition, and effectiveness

Official development assistance (ODA) has increased significantly over the past decade, nearly doubling as a share of donor gross nation income. But it has fallen short of a number of internationally agreed targets. Programmed aid for 2011–13 indicates that the growth of ODA disbursements is on track to slow in real terms and indeed shrink on a per capita basis for recipient countries.

Surprisingly, the aid directed toward agriculture, food, and nutrition—10 percent of total commitments in 2010—has not increased in response to the recent food price spikes or since the MDGs were agreed in 2000. And assistance for nutrition represents a mere fraction of these commitments (about 3 percent of total aid flows to agriculture, food, and nutrition), despite widespread evidence that improving nutrition and making gains in early childhood development are keys to meeting several MDGs and to making long-term progress in development.

Looking ahead, aid flows appear set to slow, likely reflecting the need for sharp fiscal consolidation for many large donors. Based on reported donor plans during 2011–13, disbursements for country programmable aid (accounting for roughly 60 percent of total ODA) will actually fall slightly by a real 0.2 percent a year on average.

Meeting the MDGs requires that aid flows are used as effectively as possible. In the Paris Declaration on Aid Effectiveness (2005) and the Accra Agenda for Action (2008), agreements reached at the Third and Fourth High Level Forums, the international community set out several principles and committed to specific actions under each principle, with the goal of increasing the effectiveness of the aid delivered—as well as the level of disbursements. Some progress toward greater aid effectiveness has been made. But only 1 of the Paris Declaration's 13 targets for 2010 has been met, with progress limited for the other 12. Even so, the goals and associated policy adjustments made at the forums seem to have contributed to the significant rise in aid flows.

Efforts to improve aid effectiveness are being pursued against the backdrop of fundamental changes in aid architecture. The aid agenda is shifting, with calls for stronger leadership and ownership by recipients, more harmonization and coordination among donors, and greater transparency. The donor community has dramatically expanded and become much more diverse. Many new private and public donors are coming onto the stage, among them nongovernmental organizations (including philanthropists and corporations) and a growing number of middle-income countries. The sharp rise in stakeholders highlights the way development demands an extensive set of tools and partnerships.

ODA is increasingly viewed as only one of many international activities (such as trade and investment) that support long-term sustainable development and poverty alleviation. But it remains a major instrument for development cooperation. The international aid community needs to continue to improve information-sharing and to facilitate the participation of the expanding ODA agents in setting the global development agenda—to better address the needs of the poor, including such critical issues as food and nutrition.

Note

1. Focus groups and interviews were carried out in 17 countries with respondents representing groups exposed to economic shocks, such as workers in export-oriented sectors, informal sector workers, and farmers. The research explored the extent to which people were able to remain resilient against the recent economic shocks and the means they used to do so. Data came from up to four rounds of qualitative research at sites in Bangladesh, Cambodia, Central African Republic, Ghana, Indonesia, Jamaica, Kazakhstan, Kenya, Mongolia, Philippines, Senegal, Serbia, Thailand, Ukraine, Vietnam, the Republic of Yemen, and Zambia (see chapter 1).

Goals and Targets from the Millennium Declaration

2

Achieve universal primary education

TARGET 2.A Ensure that by 2015, children everywhere, boys and girls alike, will be able to complete a full course of primary schooling

1

Eradicate extreme poverty and hunger

TARGET 1.A Halve, between 1990 and 2015, the proportion of people whose income is less than $1.25 a day

TARGET 1.B Achieve full and productive employment and decent work for all, including women and young people

TARGET 1.C Halve, between 1990 and 2015, the proportion of people who suffer from hunger

3

Promote gender equality and empower women

TARGET 3.A Eliminate gender disparity in primary and secondary education, preferably by 2005, and at all levels of education no later than 2015

4

Reduce child mortality

TARGET 4.A Reduce by two-thirds, between 1990 and 2015, the under-five mortality rate

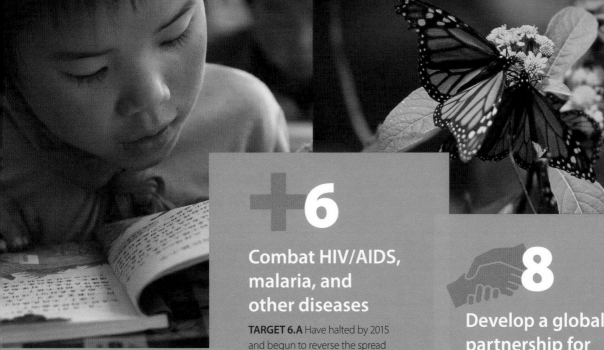

6 Combat HIV/AIDS, malaria, and other diseases

TARGET 6.A Have halted by 2015 and begun to reverse the spread of HIV/AIDS

TARGET 6.B Achieve by 2010 universal access to treatment for HIV/AIDS for all those who need it

TARGET 6.C Have halted by 2015 and begun to reverse the incidence of malaria and other major diseases

5 Improve maternal health

TARGET 5.A Reduce by three-quarters, between 1990 and 2015, the maternal mortality ratio

TARGET 5.B Achieve by 2015 universal access to reproductive health

7 Ensure environmental sustainability

TARGET 7.A Integrate the principles of sustainable development into country policies and programs and reverse the loss of environmental resources

TARGET 7.B Reduce biodiversity loss, achieving by 2010 a significant reduction in the rate of loss

TARGET 7.C Halve by 2015 the proportion of people without sustainable access to safe drinking water and basic sanitation

TARGET 7.D Have achieved a significant improvement by 2020 in the lives of at least 100 million slum dwellers

8 Develop a global partnership for development

TARGET 8.A Develop further an open, rule-based, predictable, nondiscriminatory trading and financial system (including a commitment to good governance, development, and poverty reduction, nationally and internationally)

TARGET 8.B Address the special needs of the least-developed countries (including tariff- and quota-free access for exports of the least-developed countries; enhanced debt relief for heavily indebted poor countries and cancellation of official bilateral debt; and more generous official development assistance for countries committed to reducing poverty)

TARGET 8.C Address the special needs of landlocked countries and small island developing states (through the Programme of Action for the Sustainable Development of Small Island Developing States and the outcome of the 22nd special session of the General Assembly)

TARGET 8.D Deal comprehensively with the debt problems of developing countries through national and international measures to make debt sustainable in the long term

TARGET 8.E In cooperation with pharmaceutical companies, provide access to affordable, essential drugs in developing countries

TARGET 8.F In cooperation with the private sector, make available the benefits of new technologies, especially information and communications

Source: United Nations. 2008. *Report of the Secretary-General on the Indicators for Monitoring the Millennium Development Goals.* E/CN.3/2008/29. New York.
Note: The Millennium Development Goals and targets come from the Millennium Declaration, signed by 189 countries, including 147 heads of state and government, in September 2000 (http://www.un.org/millennium/declaration/ares552e.htm) and from further agreement by member states at the 2005 World Summit (Resolution adopted by the General Assembly—A/RES/60/1). The goals and targets are interrelated and should be seen as a whole. They represent a partnership between the developed countries and the developing countries "to create an environment—at the national and global levels alike—which is conducive to development and the elimination of poverty."

Progress toward the MDGs

Global progress toward the 2015 Millennium Development Goals (MDGs) varies across targets and regions. At the global level, current estimates indicate that targets related to extreme poverty (MDG 1.a) and access to safe drinking water (MDG 7.c) have been reached (figure 1). Accordingly, the proportion of people whose income is less than $1.25 a day has decreased by at least 50 percent since 1990, when global poverty was estimated at 43.1 percent. Similarly, the proportion of people without sustainable access to safe drinking water has been halved from the 24 percent estimated for 1990.

Progress is also significant for primary completion (MDG 2.a) and gender equality in primary and secondary education (MDG 3.a). Latest available data suggest that developing countries are within 10 percentage points of the on-track trajectory (figure 1), meaning that at current trends these two development goals will likely be reached by the year 2015.

On the other hand, progress has been lagging for health-related MDGs. Global targets related to infant and maternal mortality (MDGs 4.a and 5.a), and to a lesser extent, access to basic sanitation (MDG 7.c) are significantly off-track (figure 1). Current progress in reducing by three-quarters the maternal mortality ratio roughly represents half of the required improvement needed to reach the 2015 goal.

Progress toward the 2015 goals is related to income and institutions. Nonfragile upper-middle-income countries have reached or are on track to achieve, on average, six development targets, whereas countries in fragile situation are considerably lagging behind, with only two goals achieved or on track. Nonfragile low- and lower-middle-income countries (with three and four goals, respectively, achieved or on track) have also performed better than countries in fragile situations, although not as well as upper-middle-income countries.

At the regional level, progress toward the MDGs is more diverse, although health-related targets will likely be missed in most regions (figure 2). In East Asia and Pacific the targets on extreme poverty, gender parity, and access to water and sanitation have been reached. Progress is substantial with regard to primary completion, and the goal should be achieved in the years remaining to 2015. Child and maternal mortality are the targets lagging the most.

In Europe and Central Asia the proportion of poor has been halved since 1990, and the target on access to water has been reached. Progress toward achieving universal primary education and promoting gender equality is currently on track. Increased efforts must be undertaken with regard to improving maternal health and access to basic sanitation.

Latin America and the Caribbean has already reached the targets on extreme poverty, primary completion, gender equality, and access to safe water. The region is performing better than the rest of the developing world in relation to child mortality, having achieved more than 60 percent of the progress needed to reduce under-5 mortality by two-thirds. However, Latin America and the Caribbean faces serious challenges regarding maternal mortality, as progress in this MDG has been significantly slow.

Middle East and North Africa has reached the poverty target as well as the target on access to improved sanitation facilities. The region is making fast progress toward achieving universal primary education and gender equality. Nevertheless, progress toward ensuring access to safe drinking water and eradicating maternal mortality is lagging.

South Asia has reached the target on access to safe water and will probably eliminate gender disparity in primary and secondary education by 2015. Progress has also been made with respect to primary completion and, to a lesser extent, extreme poverty reduction. Faster progress is required in terms of reducing child and maternal mortality and improving access to sanitation facilities if the region is to reach these goals by 2015.

Sub-Saharan Africa is lagging with respect to other regions and most MDGs. However, the region has achieved more than 60 percent of the progress required to reach, by 2015, goals such as gender parity, primary completion, access to safe water, and extreme poverty. As for other regions, health-related MDGs, particularly maternal mortality, require urgent attention.

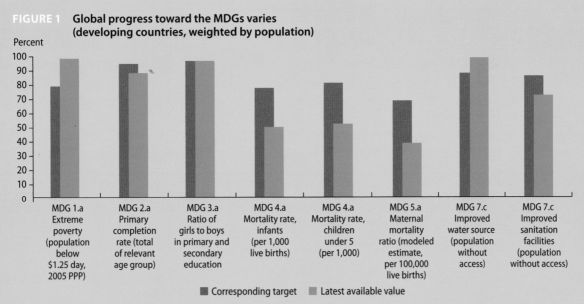

FIGURE 1 Global progress toward the MDGs varies
(developing countries, weighted by population)

■ Corresponding target ■ Latest available value

Source: World Bank staff calculations based on data from the World Development Indicators database.

Note: A value of 100 percent means that the respective MDG has been reached. "Corresponding target" indicates progress presently needed to reach the goal by 2015. "Latest available value" denotes present progress as illustrated by most recent available data: extreme poverty, 2010; primary completion rate, 2009; ratio of girls to boys in primary and secondary education, 2009; mortality rate, infants, 2010; mortality rate, children under 5, 2010; maternal mortality ratio, 2008; improved water source, 2010; improved sanitation facilities, 2008).

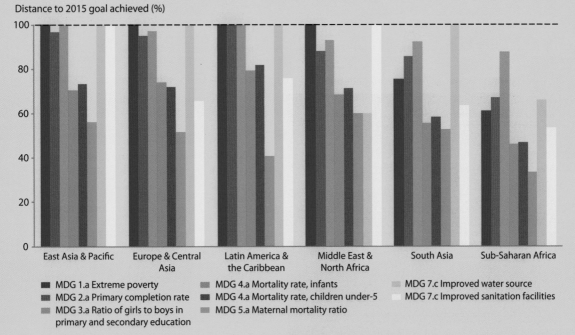

FIGURE 2 Regional progress toward the MDGs
(developing countries, weighted by population)

■ MDG 1.a Extreme poverty   ■ MDG 4.a Mortality rate, infants   ■ MDG 7.c Improved water source
■ MDG 2.a Primary completion rate   ■ MDG 4.a Mortality rate, children under-5   ■ MDG 7.c Improved sanitation facilities
■ MDG 3.a Ratio of girls to boys in primary and secondary education   ■ MDG 5.a Maternal mortality ratio

Source: World Bank staff calculations based on data from the World Development Indicators database.

Note: A value of 100 percent means that the respective MDG has been reached. Values denote present progress as illustrated by most recent available data: extreme poverty, 2010; primary completion rate, 2009; ratio of girls to boys in primary and secondary, 2009; mortality rate, infants, 2010; mortality rate, children under 5, 2010; maternal mortality ratio, 2008; improved water source, 2010; improved sanitation facilities, 2008).

Eradicating extreme poverty and hunger

Poverty and hunger remain, but fewer people live in extreme poverty. The proportion of people living on less than $1.25 a day fell from 43.1 percent in 1990 to 22.2 percent in 2008. While the food, fuel, and financial crises over the past four years have worsened the situations of vulnerable populations and slowed the rate of poverty reduction in some countries, global poverty rates kept falling. Between 2005 and 2008 both the poverty rate and the number of people living in extreme poverty fell in all six developing regions, the first time that has happened. Preliminary estimates for 2010 show that the extreme poverty rate fell further, reaching the global target of the MDGs of halving world poverty five years early. Three regions—East Asia and Pacific, Europe and Central Asia, and the Middle East and North Africa—met or exceeded the target by 2008.

Further progress is possible and likely before the 2015 target date of the MDGs, if developing countries maintain the robust growth rates achieved over much of the past decade. But even then, hundreds of millions of people will remain mired in poverty, especially in Sub-Saharan Africa and South Asia and wherever poor health and lack of education deprive people of productive employment; where environmental resources have been depleted or spoiled; and where corruption, conflict, and misgovernance waste public resources and discourage private investment.

The most rapid decline in poverty occurred in East Asia and Pacific, where extreme poverty in China fell from 60 percent in 1990 to 13 percent. In the developing world outside China, the poverty rate fell from 37 percent to 25 percent. Poverty remains widespread in Sub-Saharan Africa and South Asia, but progress in both regions has been substantial. In South Asia the poverty rate fell from 54 to 36 percent. In Sub-Saharan Africa the poverty rate fell by 4.8 percentage points to less than 50 percent between 2005 and 2008, the largest drop in Sub-Saharan Africa since international poverty rates have been computed.

In 2008 1.28 billion people lived on less than $1.25 a day. Since 1990 the number of people living in extreme poverty has fallen in all regions except Sub-Saharan Africa, where population growth exceeded the rate of poverty reduction, increasing the number of extremely poor people from 290 million in 1990 to 356 million in 2008. The largest number of poor people remain in South Asia, where 571 million people live on less than $1.25 a day, down from a peak of 641 million in 2002.

Undernourishment measures the availability of food to meet people's basic energy needs. The MDGs call for cutting the proportion of undernourished people in half, but few countries will reach that target by 2015. Rising agricultural production has kept ahead of population growth, but rising food prices and the diversion of food crops to fuel production have reversed the declining rate of undernourishment since 2004–06. The FAO estimates that in 2008 there were 739 million people without adequate daily food intake.

Rates of malnutrition have dropped substantially since 1990, but over 100 million children under age 5 remain malnourished. Only 40 countries, out of 90 with adequate data to monitor trend, are on track to reach the MDG target. Malnutrition in children often begins at birth, when poorly nourished mothers give birth to underweight babies. Malnourished children develop more slowly, enter school later, and perform less well. Programs to encourage breastfeeding and improve the diets of mothers and children can help.

FIGURE 1a Poverty rates fell sharply in the new millennium

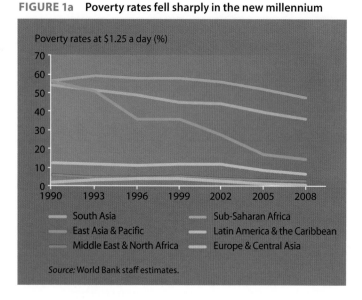

Poverty rates at $1.25 a day (%)

Source: World Bank staff estimates.

FIGURE 1b Fewer people living in extreme poverty

People living on $1.25 a day or less (millions)

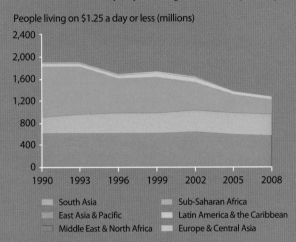

- South Asia
- East Asia & Pacific
- Middle East & North Africa
- Sub-Saharan Africa
- Latin America & the Caribbean
- Europe & Central Asia

Source: World Bank staff estimates.

FIGURE 1c Progress toward reducing undernourishment

Share of countries in region making progress (%)

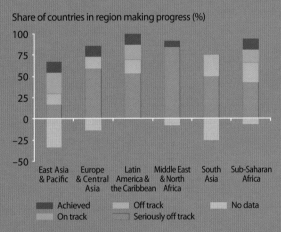

East Asia & Pacific · Europe & Central Asia · Latin America & the Caribbean · Middle East & North Africa · South Asia · Sub-Saharan Africa

- Achieved
- On track
- Off track
- Seriously off track
- No data

Source: World Bank staff estimates.

FIGURE 1d Many children remain malnourished

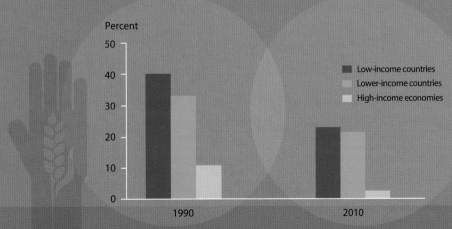

Percent

- Low-income countries
- Lower-income countries
- High-income economies

Source: World Health Organization; World Development Indicators database.
Note: "Malnourishment" is the measure of underweight children.

15

MDG 2

Achieve universal primary education

The commitment to provide primary education to every child is the oldest of the MDGs, having been set down at the first Education for All conference in Jomtien, Thailand, more than 20 years ago. This goal has been reached only in Latin America and the Caribbean, although East Asia and Pacific and Europe and Central Asia are close. Progress among the poorest countries, slow in the 1990s, has accelerated since 2000, particularly in South Asia and Sub-Saharan Africa, but the goal of full enrollment remains elusive. And even as countries approach the target, the educational demands of modern economies expand. In the 21st century, primary education will be of value only as a stepping stone toward secondary and higher education.

In 2009 87 percent of children in developing countries completed primary school. In most regions school enrollments picked up after the MDGs were promulgated in 2000, when the completion rate stood at 80 percent. Sub-Saharan Africa and South Asia, which started out farthest behind, have continued to make substantial progress but will still fall short of the goal. The Middle East and North Africa has stalled at completion rates of around 90 percent, while Europe and Central Asia and East Asia and

Pacific are within striking distance but have made little progress in the last five years.

Sixty developing countries, one-half the number of countries for which there are adequate data, have achieved or are on track to achieve the MDG target of a full course of primary schooling for all children. Twelve more will miss the 2015 deadline, but are making slow progress. That leaves at least 48 countries seriously off track, making little or no progress, 30 of them in Sub-Saharan Africa.

FIGURE 2a   The last step toward education for all

Primary school completion rate (%)

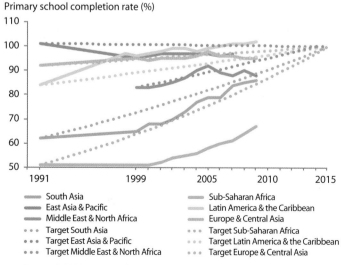

South Asia
East Asia & Pacific
Middle East & North Africa
• • • Target South Asia
• • • Target East Asia & Pacific
• • • Target Middle East & North Africa

Sub-Saharan Africa
Latin America & the Caribbean
Europe & Central Asia
• • • Target Sub-Saharan Africa
• • • Target Latin America & the Caribbean
• • • Target Europe & Central Asia

Source: UNESCO Institute of Statistics and World Development Indicators database.

FIGURE 2b   Progress toward primary education for all

Share of countries in region making progress (%)

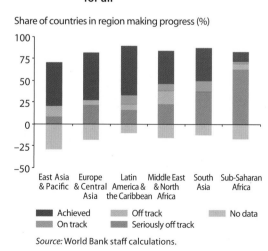

Achieved   Off track   No data
On track   Seriously off track

Source: World Bank staff calculations.

16

Promote gender equality and empower women

Women are making progress. The MDGs monitor progress toward gender equity and the empowerment of women along three dimensions: education, employment, and participation in public decision making. These are important, but there are other dimensions. Efforts are underway to improve the monitoring of women's access to financial services, entrepreneurship, migration and remittances, and violence against women. Time use surveys, for example, can do much to illuminate differences in the roles of women and men within the household and the workplace. Disaggregation of other statistical indicators by sex can also reveal patterns of disadvantage or, occasionally, advantage for women. Whatever the case, women make important contributions to economic and social development. Expanding opportunities for them in the public and private sectors is a core development strategy. And good statistics are essential for developing policies that effectively promote gender equity and increase the welfare and productivity of women.

Girls have made substantial gains in primary and secondary school enrollments. In many countries, girls'

enrollment rates outnumber boys', particularly in secondary school. But the comparison of enrollment rates obscures the underlying problem of underenrollment. Girls are still less likely to enroll in primary school or to stay in school until the end of the primary stage. In some countries the situation changes at the secondary stage. Girls who complete primary school may be more likely to stay in school, while boys drop out. In Europe and Central Asia and Latin America and the Caribbean the differences between boys' and girls' enrollments in higher education are substantial. This is an unsatisfactory path to equity. Rapid growth and poverty reduction truly requires education for all.

Substantial progress has been made toward increasing the proportion of girls enrolled in primary and secondary education. By the end of the 2009/10 school year, 96 countries had achieved equality of enrollment rates, and 7 more were on track to do so by 2015. That leaves only 27 countries off track or seriously off track, mostly low- and lower-middle-income countries in the Middle East and North Africa, South Asia, and Sub-Saharan Africa. Fourteen countries lacked adequate data to assess progress.

FIGURE 3a **Increasing participation by girls at all levels of education**

Ratio of girls' to boys' enrollment rate, 2009 (%)

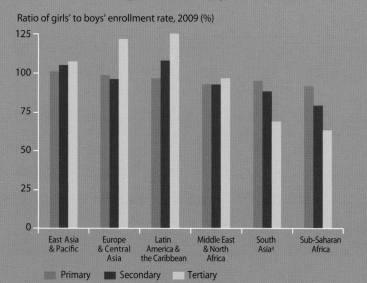

FIGURE 3b **Progress toward gender equality in primary and secondary education**

Share of countries in region making progress (%)

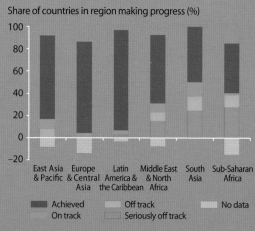

Source: UNESCO Institute of Statistics and World Development Indicators (WDI) database.
a. Data for primary and tertiary enrollment are from 2008.

Reduce child mortality

Deaths in children under age 5 have been declining since 1990. In 2006, for the first time, the number of children who died before their fifth birthday fell below 10 million. In developing countries the mortality rate has declined from 98 per 1,000 in 1990 to 63 in 2010. Still, progress toward the MDG target of a two-thirds reduction has been slow. In Sub-Saharan Africa, one child in 8 dies before their fifth birthday. The odds are somewhat better in South Asia, where one child in 15 dies before their fifth birthday. Even in regions with relatively low mortality rates, such as Latin America and the Caribbean and Europe and Central Asia, slow improvements leave most countries well short of the MDG target.

Thirst-six developing countries have achieved or are now on track to achieve the target of a two-thirds reduction in under-five mortality rates.

Distribution of progress

In 1990 the under-5 mortality rate in Niger stood at 311 per 1,000, the worst in the world. In the same year, Seychelles, with an under-5 mortality rate of 16, was the best in Sub-Saharan Africa. How have they fared since? In the 20 years since the MDG baseline, Niger's mortality rate fell by 168 points, the greatest in the region, while Seychelles' fell by 3 points. In proportional terms, Niger experienced a 54 percent reduction—second greatest in the region—and Seychelles a 16 percent reduction. Both fall short of the MDG target, but Niger, starting in last place, has progressed somewhat faster. Has this been the general rule? Figure 4c shows the 1990 under-5 mortality rates for all low- and middle-income countries in Sub-Saharan Africa in 1990 and the improvement to 2010. Only one country, Zimbabwe, moved backward. Two countries, Malawi and Madagascar are on track to achieve the MDG target. Several others, including Niger, Eritrea, and Tanzania are close. The downward sloping regression line (white dots) shows the expected reduction in mortality rates, given countries' starting position. On average, countries starting in worse positions in Sub-Saharan Africa have done better, possibly because large scale vaccination programs, the introduction of treated bed nets as a malaria preventative, and campaigns to encourage exclusive breastfeeding have been able to reach a large number of people, even in poor countries. But as Figure 4c also reveals, the experience has been highly mixed: Conflict-affected countries, like Somalia and the Democratic Republic of the Congo, have made almost no progress, while similarly situated countries such as Zambia and Uganda have done much better.

FIGURE 4a   Still far to go

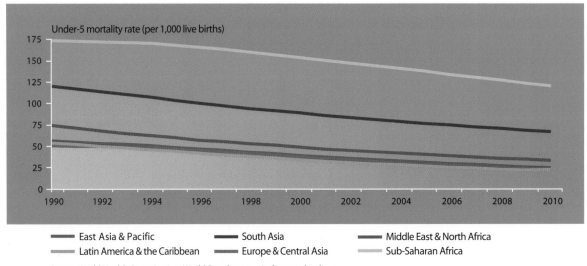

Under-5 mortality rate (per 1,000 live births)

East Asia & Pacific   South Asia   Middle East & North Africa
Latin America & the Caribbean   Europe & Central Asia   Sub-Saharan Africa

Source: World Health Organization; World Development Indicators database.

FIGURE 4b   Most deaths happen in the first year

Child deaths, 2010 (thousands)

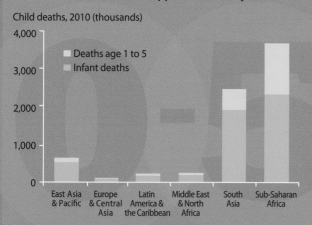

Deaths age 1 to 5
Infant deaths

| East Asia & Pacific | Europe & Central Asia | Latin America & the Caribbean | Middle East & North Africa | South Asia | Sub-Saharan Africa |

Source: World Health Organization; World Development Indicators database.

FIGURE 4d   Progress toward reducing child mortality

Share of countries in region making progress (%)

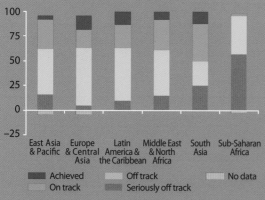

East Asia & Pacific   Europe & Central Asia   Latin America & the Caribbean   Middle East & North Africa   South Asia   Sub-Saharan Africa

Achieved   Off track   No data
On track   Seriously off track

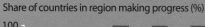

Source: World Bank staff calculations.

FIGURE 4c   For some, better than expected improvements

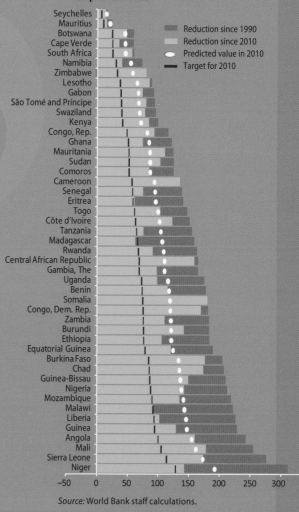

Reduction since 1990
Reduction since 2010
Predicted value in 2010
Target for 2010

Source: World Bank staff calculations.

19

Reduce maternal mortality

An estimated 358,000 maternal deaths occurred worldwide in 2008, a 34 percent decrease since 1990. Most maternal deaths occurred in developing countries. What makes maternal mortality such a compelling problem is that it strikes young women experiencing a natural life event. They die because they are poor. Malnourished. Weakened by disease. Exposed to multiple pregnancies. And they die because they lack access to trained health care workers and modern medical facilities. Death in childbirth is a rare event in rich countries, where there are typically fewer than 15 maternal deaths for every 100,000 live births, an average that has remained essentially constant for the past 18 years. And because women in poor countries have more children, their lifetime risk of maternal death may be more than 200 times greater than for women in Western Europe and North America.

Reducing maternal mortality requires a comprehensive approach to women's reproductive health, starting with family planning and access to contraception. Many health problems among pregnant women are preventable or treatable through visits with trained health workers before childbirth. Good nutrition, vaccinations, and treatment of infections can improve outcomes for mother and child. Skilled attendants at time of delivery and access to hospital treatments are essential for dealing with life-threatening emergencies such as severe bleeding and hypertensive disorders.

About half of all maternal deaths occur in Sub-Saharan Africa and a third in South Asia. but mothers face substantial risks in other regions as well. Among fragile and conflict-affected states, the mortality ratio may be many times higher.

Progress in reducing maternal mortality ratios has been slow, far slower than imagined by the MDG target of a 75 percent reduction from 1990 levels. Accurate measurement of maternal mortality is difficult, requiring accurate reporting of vital events and specialized surveys. Recent efforts by statisticians have improved estimates, but for many countries the need for improved monitoring of maternal health will continue long past 2015.

Women who give birth at an early age are likely to bear more children and are at greater risk of death or serious complications from pregnancies. In many developing countries, the number of women ages 15–19 is still increasing. Preventing unintended pregnancies and delaying childbirth among young women increase the chances of their attending school and eventually obtaining paid employment.

Having skilled health workers present for deliveries is key to reducing maternal mortality. In many places women have only untrained caregivers or family members to attend them during childbirth. Skilled health workers are trained to give necessary care before, during, and after delivery; they can conduct deliveries on their own, summon additional help in emergencies, and provide care for newborns.

FIGURE 5a   Maternal mortality ratios have been falling but large regional differences persist

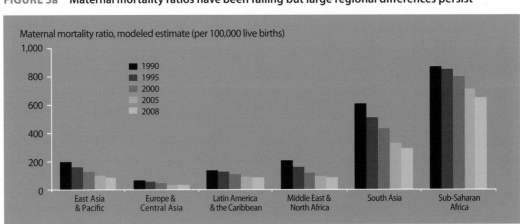

Source: World Health Organization; World Development Indicators database.

FIGURE 5b Progress in reducing maternal mortality

Share of countries in region making progress (%)

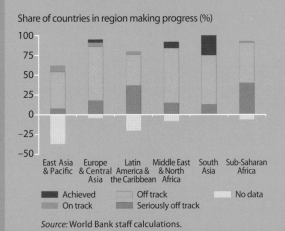

Legend:
- Achieved
- On track
- Off track
- Seriously off track
- No data

Source: World Bank staff calculations.

FIGURE 5d Fewer young women giving birth

Adolescent fertility rate (births per 1,000 women ages 15–19)

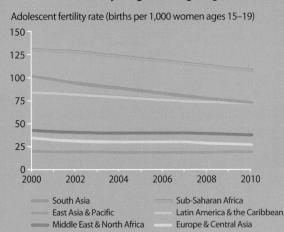

Legend:
- South Asia
- East Asia & Pacific
- Middle East & North Africa
- Sub-Saharan Africa
- Latin America & the Caribbean
- Europe & Central Asia

Source: World Health Organization; World Development Indicators database.

FIGURE 5c Help for mothers

Births attended by skilled health staff, 2010 (% of total)

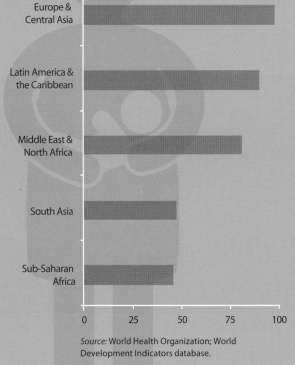

Source: World Health Organization; World Development Indicators database.

MDG 6

Combat HIV/AIDS, malaria, and other diseases

Epidemic diseases exact a huge toll in human suffering and lost opportunities for development. Poverty, armed conflict, and natural disasters contribute to the spread of disease and are made worse by it. In Africa the spread of HIV/AIDS has reversed decades of improvement in life expectancy and left millions of children orphaned. It is draining the supply of teachers and eroding the quality of education.

There are 300 million to 500 million cases of malaria each year, leading to more than 1 million deaths. Nearly all the cases occur in Sub-Saharan Africa, and most deaths from malaria are among children younger than 5.

Tuberculosis kills some 2 million people a year, most of them 15–45 years old. The disease, once controlled by antibiotics, is spreading again because of the emergence of drug-resistant strains. People living with HIV/AIDS, which reduces resistance to tuberculosis, are particularly vulnerable as are refugees, displaced persons, and prisoners living in close quarters and unsanitary conditions.

Sub-Saharan Africa remains at the center of the HIV/AIDS epidemic, but the proportion of adults living with AIDS has begun to fall even as the survival rate of those with access to antiretroviral drugs has increased. In Africa 58 percent of the adults with HIV/AIDS are women. The region with the next-highest prevalence rate is Latin America and the Caribbean, where 0.5 percent of adults are infected.

In 2009 between 31 million and 33 million people were living with HIV/AIDS. Of these approximately 1.5 million were under the age of 15. Another 16.9 million children, of which 14.8 million live in Sub-Saharan Africa, lost one or both parents to AIDS. By the end of 2009, 5.25 million people were receiving antiretroviral drugs, representing 36 percent of the population for which the World Health Organization recommends treatment.

The MDGs call for halting and then reversing the spread of HIV/AIDS by 2015. This progress assessment is based on prevalence rates for adults ages 15–49. Countries with declining prevalence rates since 2005 are assessed to have halted the epidemic; those with prevalence rates less than their earliest measured rate have reversed the epidemic. Countries with prevalence rates of less than 0.2 percent were considered to be stable.

Malaria is endemic in most tropical and subtropical regions, but 90 percent of the malaria deaths occur in Sub-Saharan Africa. Those most severely affected are children under age 5. Even those who survive malaria do not escape unharmed. Repeated episodes of fever and anemia take a toll on their mental and physical development. Insecticide-treated bed nets have proved to be an effective preventative. Their use has grown rapidly. Between 2008 and 2010, 290 million nets were distributed in Sub-Saharan Africa, but coverage remains uneven. In some countries with large numbers of reported cases, use of bed nets for children remains at less than 20 percent.

FIGURE 6a   Bringing HIV/AIDS under control

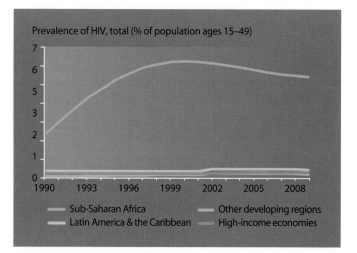

Source: World Health Organization/UNAIDS; World Development Indicators database.

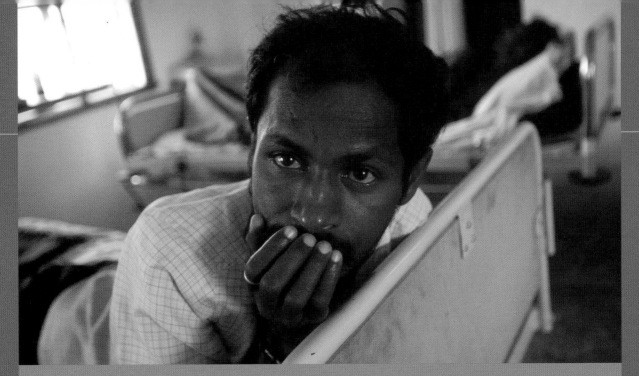

FIGURE 6b Millions of people still afflicted with HIV/AIDS

Adults and children living with HIV, 2009 (millions)

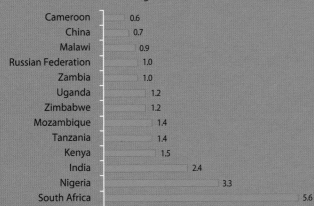

Cameroon	0.6
China	0.7
Malawi	0.9
Russian Federation	1.0
Zambia	1.0
Uganda	1.2
Zimbabwe	1.2
Mozambique	1.4
Tanzania	1.4
Kenya	1.5
India	2.4
Nigeria	3.3
South Africa	5.6

Source: World Health Organization/UNAIDS; World Development Indicators database.

FIGURE 6d Progress toward reversing the HIV/AIDS epidemic

Share of countries in region making progress (%)

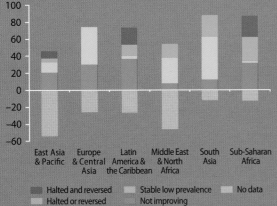

Halted and reversed — Stable low prevalence — No data
Halted or reversed — Not improving

Source: World Bank staff calculations.

FIGURE 6c Protecting children from malaria

Use of insecticide-treated bed nets
(% of under-5 population)   2008 Notified cases (per 100,000)

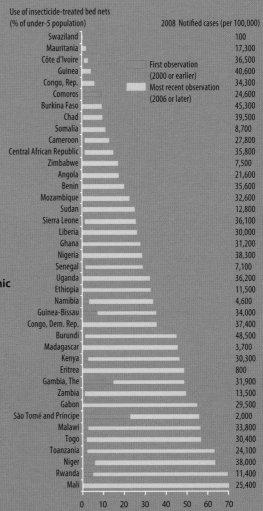

First observation (2000 or earlier)
Most recent observation (2006 or later)

Swaziland	100
Mauritania	17,300
Côte d'Ivoire	36,500
Guinea	40,600
Congo, Rep.	34,300
Comoros	24,600
Burkina Faso	45,300
Chad	39,500
Somalia	8,700
Cameroon	27,800
Central African Republic	35,800
Zimbabwe	7,500
Angola	21,600
Benin	35,600
Mozambique	32,600
Sudan	12,800
Sierra Leone	36,100
Liberia	30,000
Ghana	31,200
Nigeria	38,300
Senegal	7,100
Uganda	36,200
Ethiopia	11,500
Namibia	4,600
Guinea-Bissau	34,000
Congo, Dem. Rep.	37,400
Burundi	48,500
Madagascar	3,700
Kenya	30,300
Eritrea	800
Gambia, The	31,900
Zambia	13,500
Gabon	29,500
São Tomé and Príncipe	2,000
Malawi	33,800
Togo	30,400
Toanzania	24,100
Niger	38,000
Rwanda	11,400
Mali	25,400

0   10   20   30   40   50   60   70

Source: World Health Organization; World Development Indicators database.

23

Ensure environmental sustainability

Sustainable development can be ensured only by protecting the environment and using its resources wisely. Poor people, often dependent on natural resources for their livelihood, are the most affected by environmental degradation and natural disasters (fires, storms, earthquakes)—the effects of which are worsened by environmental mismanagement. Poor people also suffer from shortcomings in the built environment; whether in urban or rural areas, they are more likely to live in substandard housing, to lack basic services, and to be exposed to unhealthy living conditions.

Most countries have adopted principles of sustainable development and have agreed to international accords on protecting the environment. But the failure to reach a comprehensive agreement on limiting greenhouse gas emissions leaves billions of people and future generations vulnerable to the impacts of climate change. Growing populations put more pressure on marginal lands and expose more people to hazardous conditions that will be exacerbated by global warming.

Annual emissions of carbon dioxide reached 32 million metric tons in 2008 and are still rising. High-income economies remain the largest emitters, but the rapidly growing upper-middle-income countries are not far behind. Measured by emissions per capita, however, emissions by high-income economies are more than three times as high as the average of low- and middle-income countries.

Loss of forests threatens the livelihood of poor people, destroys habitats that harbor biodiversity, and eliminates an important carbon sink that helps to moderate the climate. Net losses since 1990 have been substantial, especially in Latin America and the Caribbean and Sub-Saharan Africa, and these losses are only partially compensated by increases in Asia and high-income economies.

The MDGs call for halving the proportion of the population without access to improved sanitation and water sources by 2015. As of 2010, 2.7 billion people still lacked access to improved sanitation, and more than 1 billion people practiced open defecation, posing enormous health risks. At the present pace only 37 countries are likely to reach the target—a pickup of 2 since the last measurement in 2008. East Asia and Pacific and Middle East and North Africa are the only developing regions on track to reach the target by 2015.

In 1990 more than 1 billion people lacked access to drinking water from a convenient, protected source, but the situation is improving. The proportion of people in developing countries with access to an improved water source increased from 71 percent in 1990 to 86 percent in 2008. The MDG target is to reduce by half the proportion of people without access to an improved water source. Seventy-three countries have reached or are on track to reach the target. At this rate, only the developing regions of the Middle East and North Africa and Sub-Saharan Africa will fall short.

In 1990, 63 percent of the people living in low- and middle-income countries lacked access to a flush toilet or other form of improved sanitation. By 2010 the access rate had improved by 19 percentage points to 44 percent. The situation is worse in rural areas, where 57 percent of the population lack access to improved sanitation. The large urban-rural disparity, especially in Sub-Saharan Africa and South Asia, is the principal reason the sanitation target of the MDGs will not be achieved.

FIGURE 7a   Carbon dioxide emissions continue to rise

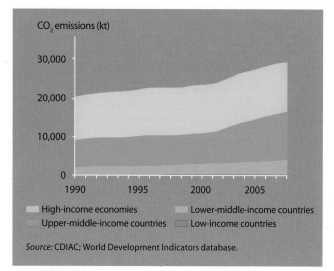

Source: CDIAC; World Development Indicators database.

FIGURE 7b   Forest losses and gains

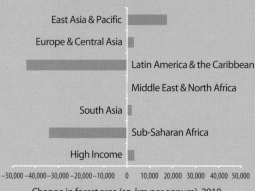

East Asia & Pacific

Europe & Central Asia

Latin America & the Caribbean

Middle East & North Africa

South Asia

Sub-Saharan Africa

High Income

−50,000 −40,000 −30,000 −20,000 −10,000 0 10,000 20,000 30,000 40,000 50,000

Change in forest area (sq. km per annum), 2010

Source: FAO; World Development Indicators database.

FIGURE 7c   Progress toward improved sanitation

Share of countries in region making progress (%)

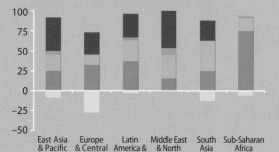

East Asia & Pacific | Europe & Central Asia | Latin America & the Caribbean | Middle East & North Africa | South Asia | Sub-Saharan Africa

- Achieved
- On track
- Off track
- Seriously off track
- No data

Source: World Bank staff calculations.

FIGURE 7d   Many still lack access to sanitation

Share of population with access to improved sanitation, 2008 (%)

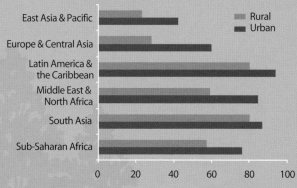

East Asia & Pacific

Europe & Central Asia

Latin America & the Caribbean

Middle East & North Africa

South Asia

Sub-Saharan Africa

Rural
Urban

0   20   40   60   80   100

Source: World Health Organization; World Development Indicators database.

FIGURE 7e   Progress toward improved water sources

Share of countries in region making progress (%)

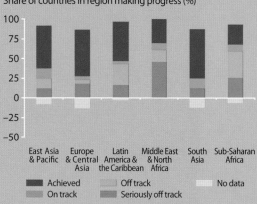

East Asia & Pacific | Europe & Central Asia | Latin America & the Caribbean | Middle East & North Africa | South Asia | Sub-Saharan Africa

- Achieved
- On track
- Off track
- Seriously off track
- No data

Source: World Bank staff calculations.

Develop a global partnership for development

The eighth and final goal distinguishes the MDGs from previous sets of resolutions and targeted programs. It recognizes the multidimensional nature of development and the need for wealthy countries and developing countries to work together to create an environment in which rapid, sustainable development is possible. Following the 2002 Millennium Summit in Monterrey, Mexico, world leaders agreed to provide financing for development through a coherent process that recognized the need for domestic as well as international resources. Subsequent high-level meetings expanded on these commitments. Along with increased aid flows and debt relief for the poorest, highly indebted countries, MDG 8 recognizes the need to reduce barriers to trade and to share the benefits of new medical and communication technologies. MDG 8 also reminds us that development challenges differ for large and small countries and for those that are landlocked or isolated by large expanses of ocean. Building and sustaining a partnership is an ongoing process that does not stop on a specific date or when a target is reached. However it is measured, a strong commitment to partnership should be the continuing legacy of the MDGs.

The financial crisis that began in 2008 and fiscal austerity in many high-income economies have threatened to undermine commitments to increase official development assistance. So far, leading donors have maintained their level of effort. Total disbursements by members of the OECD Development Assistance Committee reached $130 billion in 2010, a real increase of 4.3 percent over 2008.

OECD countries (which include some upper-middle-income countries such as Mexico and Chile) spend more on support to domestic agricultural producers than they do on official development assistance. In 2010 the OECD producer support estimate stood at $227 billion, down by about 10 percent from the previous three years.

The growth of fixed-line phone systems has peaked in high-income economies and will never achieve the same level of use in developing countries, where mobile cellular subscriptions continue to grow at a rapid pace. In high-income economies, with more than one subscription per person, the pace of growth appears to be slowing.

Growing economies, better debt management, and debt relief for the poorest countries have allowed developing countries to substantially reduce their debt burdens. Despite the financial crisis, which caused the global economy to contract by 2.3 percent in 2009, debt service ratios continued to fall in most developing regions.

FIGURE 8a   Most donors have maintained their aid levels

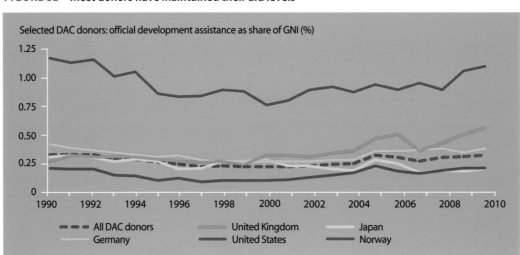

Source: Organisation for Economic Co-operation and Development

FIGURE 8b   But domestic subsidies to agricultural are greater

Selected DAC donors agricultural support as share of GDP (%)

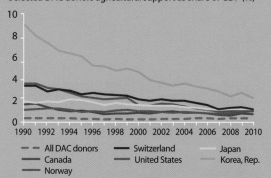

- – – All DAC donors
- —— Switzerland
- —— Japan
- —— Canada
- —— United States
- —— Korea, Rep.
- —— Norway

Source: Organisation for Economic Co-operation and Development.

FIGURE 8d   Debt service burdens have been falling

Total debt service (% of exports of goods, services, and income)

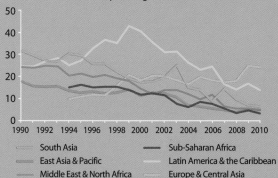

- ········ South Asia
- ——— Sub-Saharan Africa
- ――― East Asia & Pacific
- ——— Latin America & the Caribbean
- ——— Middle East & North Africa
- ········ Europe & Central Asia

Source: International Telecommunication Union and World Development Indicators database.

FIGURE 8c   Cellular phones are connecting developing countries

Fixed line and mobile cellular subscriptions (per 100 people)

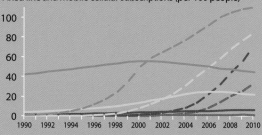

- —— Low-income country telephone lines
- —— Lower-middle-income country telephone lines
- —— Upper-middle-income country telephone lines
- —— High-income economy telephone lines
- – – – Low-income country mobile subscriptions
- – – – Lower-middle-income country mobile subscriptions
- – – – Upper-middle-income country mobile subscriptions
- – – – High-income economy mobile subscriptions

Source: World Development Indicators database.

Poverty and Food Price Developments

Summary and main messages

The food price spikes have prevented millions of people from escaping extreme poverty. The record prices in 2008 kept or pushed 105 million people below the poverty line in the short run. They hit urban poor and female-headed households hardest. While food prices dropped sharply in 2009 with the financial crisis, they quickly rebounded and by early 2011 were almost back to 2008 levels. Sudden, unexpected increases in food prices impose particularly severe hardship on many households because they need time to adjust to higher prices. The large, initial impact on poverty of a rise in food prices tends to decline over time as production increases and the income of the poor in rural areas rises, but it is usually not large enough to offset the initial negative impact on poverty in the short run.

The factors that caused the price spikes also have the potential to make prices more volatile and thus less predictable. Biofuel mandates, which have boosted demand for grains, despite slowing demand for food globally, have reduced the price elasticity of demand for grains. Sharp increases in fertilizer prices, linked to energy prices, have made production costs more volatile and, to

the extent that higher prices have reduced the use of fertilizers, have made yields less stable. Adverse weather patterns also have become more frequent and more variable. Low global stocks have contributed to price volatility at time of production shortfalls. Moreover, trade interventions meant to stabilize domestic prices often have had the adverse effect and increased price volatility globally.

The challenges differ across countries. Food price increases have different effects on a country's current account depending on whether the country is a net importer or net exporter, while the impact on a country's fiscal position depends on subsidy programs and other market interventions. In addition, the extent and speed of transmission of changes in international food prices during the recent price spikes to domestic prices has varied considerably across countries. Transmission of prices has been limited in countries that impose trade barriers and have poor infrastructure. This isolates domestic from international markets and potentially raises price volatility in domestic markets. Trade restrictions, price controls, and rationing can limit the rise in domestic food prices in response to international price spikes, but at the cost of eroding producer incentives and, in the case of export bans, perhaps encouraging responses by exporters that could increase international prices. A more efficient and sustainable response to international food price spikes would permit domestic prices to rise while increasing assistance to the poor.

Characteristics of each country determine the most appropriate policy mix for addressing the implications of higher and more volatile food prices, although the content of the chosen policies will not differ greatly among countries. The chosen policy mix at the country level depends critically on how much of a country's food needs to be imported, how much of their income the poor spend on food, the socioeconomic characteristics of the poor affected, and the political environment. It depends equally on a country's integration with regional and world markets, on its level of productivity compared with what is achievable, and on its government's capacity to target the poor and vulnerable through mitigating interventions, which vitally depends on the adoption of such programs before a crisis. In addition, the government's ability to raise public expenditures or provide tax incentives in response to a food price shock without jeopardizing fiscal sustainability depends on initial macroeconomic conditions.

In the long term, the policy mix needs to address the main bottlenecks to the functioning of the domestic food markets and profitability of farmers. This would include the use of technological innovations to improve productivity. Over the long term, policies that would limit the average rise in food prices, without undermining farmer profitability, include promoting increased yields through research, extension, and improved water management; improving the efficiency of land markets and strengthening property rights; using more efficient technologies for producing biofuels; increasing farmers' access to efficient tools to manage risk; and increasing the integration of domestic markets with world markets. Policies that would limit food price volatility include the development of weather-tolerant grain varieties, increases in the size and improvements in the management of stocks, the opening markets to trade, and improvements in market transparency.

An increase in yields is needed, especially in Sub-Saharan Africa. Yields there are well below levels achieved in other parts of the world and well below what is achievable in Sub-Saharan Africa. At the same time, population growth remains high and Africa has become increasingly dependent on food imports. Increased public and private investment, better water management, and improved farming practices to more fully exploit existing technology, as well as further research, are essential to raise yields. It is also crucial to improve the trade infrastructure, to enable more trade within Africa (World Bank 2009). Raising productivity could have a substantial impact on prices and income of farmers, lowering rural poverty and making food more affordable for the urban vulnerable and poor.

Evidence in the GMR 2011 pointed to the critical role of strong economic growth and a stable macroeconomic environment in progressing toward the MDGs. Seen in this context, the strong economic performance of emerging and developing countries in the past several years and their resilience in the face of the global financial crisis are major accomplishments. However, a weaker, more uncertain global economy in 2012, combined with still-high food prices, may pose new challenges and complicate emerging and developing countries' quest to further reduce poverty and hunger. Developing countries coped well with the recent global downturn but face the current global economic environment with depleted policy buffers. Among possible risks to the outlook is a further sharp slowdown in global growth and a new or extended spike in food prices. Should such risks materialize, possible responses must be directed toward protecting the most vulnerable and poor people within a stable and sustainable macroeconomic framework.

Rising food prices have prevented millions of people from escaping poverty

Agricultural prices in 2011 exceeded their 2008 peaks by 17 percent. Food prices increased 92 percent in nominal terms and

FIGURE 1.1   **Food, grain, agricultural, and energy price developments (in nominal and real terms)**

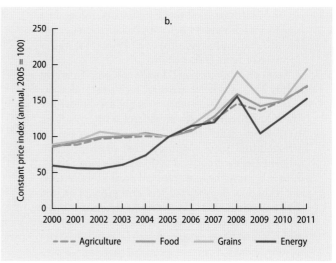

Source: World Development Indicators database.

57 percent in real terms from December 2005 to January 2012 (figure 1.1). The World Bank Agriculture Price Index peaked in February 2011, exceeding levels reached in 2008. The 2010–11 international price increases were more widespread across agricultural commodities than in 2008, when they were mainly concentrated in grain crops.[1] Since June 2010 agricultural price increases have been broad-based, affecting sugar, edible oils, beverages, animal products, and raw materials such as cotton.

High and volatile food prices can hurt food security. Large, sudden, and particularly unexpected food price increases make it difficult for households to adjust—eroding consumer purchasing power, reducing calorie intake and nutrition, and pushing more people into poverty and hunger. Overall impacts depend on the proportions of households that are net buyers and households that sell surplus production (net sellers). Net buyers will see their purchasing power decrease. Because the poor spend much of their income on food (50–70 percent), they bear a disproportionate burden in adjusting to high food prices. This is especially true for poor urban households and those headed by women, who typically spend more than half their incomes on food and are more likely to curtail consumption in the face of higher prices. At the same time,

supply shocks such as droughts can seriously derail food consumption and lead to all-out famine (box 1.1).

Qualitative survey-based research shows that responses of poor people in 13 countries to global shocks lead to severe indirect impacts.[2] Poor people have experienced a series of global shocks in recent years, from the spikes in fuel and food prices, to the economic contraction that started in 2008, while droughts have exacerbated problems

BOX 1.1   Crisis in the Horn of Africa

Below average rainfall since 2010, compounded by rising and more volatile food and fuel prices, protectionist policies, political instability and conflict, and deteriorating conditions in refugee camps have exacerbated the food crisis in the Horn of Africa. Nearly 13 million people in Djibouti, Ethiopia, Kenya, and Somalia face food insecurity; famine afflicts about 4 million people in Somalia.

Higher food prices and malnutrition remain severe problems in all of these countries
In **Somalia** domestic supply appears to cover only 15–20 percent of demand,[a] local grain prices have more than doubled since June 2010 in some areas, and continued instability is driving refugee flows to neighboring countries. According to the Food Security and Nutrition Analysis Unit-Somalia, recent data suggest that around 34 percent of children under age five are malnourished, of whom 40 percent suffer from severe acute malnutrition.

Food price inflation in **Ethiopia** reached 47 percent in July 2011, and some areas are facing exceptionally harsh conditions: wasting among children under age five in the south and southeast regions ranges from 10 to 22 percent. However, the number of people affected and total economic cost of the current drought are low compared with previous food crises,[b] in part because the most affected areas account for a small share of domestic agricultural production and livestock population. The World Bank and the International Monetary Fund (2010) estimates that the drought could reduce gross domestic product (GDP) by only about 0.5 percent, provided that rainfall conditions improve.

The price of maize in **Kenya** doubled in the year ending October 2011. Livestock is the main source of livelihood in the drought-affected areas and accounts for about 5 percent of total GDP. Estimated livestock mortality as a result of the drought is about 10–15 percent above normal in the affected areas, equivalent to 5 percent of Kenya's livestock population.[c] The Dadaab camp for Somali refugees has faced a difficult security situation. Overall, the direct negative impact of the drought is estimated at approximately 0.2 percent of GDP.

A fifth of **Djibouti's** population is in need of food relief. Low rainfall in the northwest and southeast has kept food prices high and exacerbated food insecurity

among pastoralists, while in urban areas high food prices and unemployment have increased poor households' dependence on food aid. Moderate malnutrition among children under five tripled in poor urban areas between May 2010 and May 2011, affecting approximately 26,000 children.[d]

The international development community is responding to the crisis but more funds are needed
As of December 16, 2011, funding coverage for humanitarian assistance in the four drought-affected countries in the Horn of Africa was estimated at 79 percent of need.[e] Increased support is particularly needed for humanitarian assistance in Djibouti and refugee-related requirements in Ethiopia.

The World Bank's International Development Association, the donor-funded Global Facility for Disaster Reduction and Recovery, and the State and Peace Building Fund are making available $1.88 billion to address short-term crisis mitigation and long-term development objectives. A total of $288 million has been allocated for the rapid response phase, which will provide health services (health screenings and nutrition schemes) and safety net programs (cash for work and cash transfer programs) through early 2012. The economic recovery phase will provide $384 million over a two-year period, to support agriculture and livestock production by improving land management and irrigation. The final, drought resilience phase will allocate $1.2 billion to drought-resilient agriculture, risk financing, resilience planning and strengthening social safety nets.[f]

a. Famine Early Warning System Network. 2011. "Special Brief: Market Functioning in Southern Somalia." U.S. Agency for International Development, Washington, DC (July 28).

b. World Bank. 2011. "Impact of the Drought and the Rise in Food Prices: Ethiopia," Country Assessment, Washington, DC (November).

c. World Bank. 2011. "The Drought and Food Crisis in the Horn of Africa: Impacts and Proposed Policy Responses for Kenya." PREM Economic Premise 71, Washington, DC (November).

d. United Nations Office for the Coordination of Humanitarian Affairs, 2012; UNICEF. 2011. "Feeding Centers Aim to Alleviate Chronic Malnutrition in Drought-Affected Djibouti." *At a Glance: Djibouti*, August 18.

e. United Nations Office for the Coordination of Humanitarian Affairs, 2012, p. 6.

f. World Bank. 2011. "Response Plan, Drought in the Horn of Africa." September 10.

BOX 1.2   How rising food prices affect the citizens of Dar es Salaam

In 2011 Tanzanians were hit by substantial increases in commodity prices. The country's inflation rate rose throughout the year, reaching 18 percent by December. How did rising food prices affect the citizens of Dar es Salaam, and did they change their consumption patterns? The World Bank worked with the NGO Twaweza to use mobile phones to survey households on their perceptions.

The number of low-income households that could afford three meals a day has fallen by about 20 percent since the end of 2010. The reported consumption of a number of food types also decreased for individual households (box figure).

High inflation and rising commodity prices were also reflected in citizens' general assessment of their economic situation. In 2010 about half the respondents (51.3 percent) were negative about their economic situation. In 2011 the proportion rose to nearly three in four (72.5 percent). And the percentage of citizens who thought Tanzania's economic situation was bad or very bad rose from 65.7 percent in 2010 to 85.7 percent in the current study.

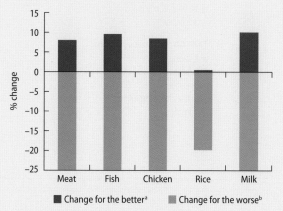

Changes in household consumption patterns

■ Change for the better[a]   ■ Change for the worse[b]

a. Percentage of households that reported this type of food as part of a typical family meal in the current study, but not in the baseline.
b. Percentage of households that reported this type of food as part of a typical family meal in the baseline, but not in the current study.

at the local level (Heltberg, Hossain, and Reva 2012, forthcoming). The shocks often resulted in severe hardships, and responses led to second-order impacts. Less nutritious diets caused malnourishment and made people more susceptible to health shocks. The sudden influx of workers into the informal economy lowered earnings. Such extreme hardship can even lead to criminal activities, eroding trust and cohesion in communities (box 1.2).

Reducing the quality of food and the number of meals was one of the most common responses, often the first to be used, in study sites in all countries surveyed (table 1.1). In addition, reducing nonfood consumption, working more hours, and diversifying sources of income (say, by entering a new informal occupation) were common nearly everywhere. Migration, sometimes as reverse migration to the home area, was also a fairly common response to the food price spikes.

Asset sales were common in many sites. Loans from family, friends, and moneylenders were also important in many countries. Inability to service microfinance and moneylender debts was a major source of distress in some East and South Asian countries, where many people had to borrow at very high interest rates to service microfinance debts or live in fear of creditors taking possession of their property. Collecting food and fuel from common property natural resources was important only in some low-income countries.

Some of these hardships (sales of productive assets, forgone education, and health care) will have long-lasting consequences and impede people's ability to recover. And coping with economic crises has eroded the savings and asset base of many households, leaving them with few resources to cope with other shocks. Continuing high and volatile global food prices are thus a major source of concern.

Many parents sought to protect children's food consumption and schooling, with adult household members saving on the quantity

TABLE 1.1   Common coping responses to food, fuel, and financial crises in 13 countries

Behavior-based responses	Number of countries	Asset-based responses	Number of countries
Reduce the quality and quantity of food	13	Sell assets	8
Reduce nonfood expenditures	13	Loan from formal lender	2
Stop primary or secondary education	6	Microfinance loan	2
Stop higher education	2	Loan from family/friends	7
Work more	12	Loan from moneylender	4
		Use common property natural resources for fuel and food	4
Take up illicit occupations:			
Sex work	2	**Assistance-based responses**	
Drug dealing	2		
Crime/theft	10	**Source of assistance:**	
Income diversification	9	Government	4
Migration	6	Nongovernmental organization	4
		Religious organization	5
		Mutual solidarity group	7
		Relatives	13
		Friends and neighbors	11

Source: Heltberg, Hossain, and Reva 2012, forthcoming.

and quality of food to ensure that children had proper diets. Yet, there were many instances of erratic attendance and school withdrawals because of the need for children to contribute to household income or because education costs had become prohibitive. But, on the whole, the impacts on schooling were more muted than expected. The cost of education, the distance to schools, and the availability of school feeding influenced whether children stayed in school.

Food price spikes have an immediate impact on progress toward eradicating extreme (income) poverty. The international food price spike of 2007–08 is estimated to have kept or pushed 105 million into poverty, and that of 2010–11 by 48.6 million people in the short run (box 1.3). Poverty typically increases initially with higher food prices, because the supply response to rising prices takes time to materialize and many poor (farm) households are net food buyers, so higher food prices lower their real incomes.

Once farm wages and farm production adjust, the impact of higher food prices on poverty is greatly ameliorated. Higher farm wages and supply responses by both smallholder and large commercial farmers dampen the impact on poverty, but it is usually not sufficient to fully offset the negative short-term impact on poverty. Some net buyers become net sellers, and higher farm wage income can offset some or all of the negative impact of higher food prices on the incomes of net consumers (see box 1.3). For this positive effect to occur, prices need to remain relatively stable and at their elevated levels, so that farmers are comfortable shifting to more profitable crops and expanding production. Hence, increased food price volatility could derail this positive development.

The impact of higher food prices differs across socioeconomic groups. Urban, nonfarm, and female-headed households are affected the most in the short term (see figure 1.2 and box 1.4). In the short term, the poverty impact of a doubling of food prices is on average 16.7 percent larger in female-headed households than in male-headed households. Short-term changes in poverty are likely to be 2.4 times higher for nonfarm households than for farm households, and the poverty impact in urban areas in the short term is likely to be 44.3 percent higher than in rural areas. The short-term effect is reduced when wages increase and farmers switch to those products that increase their profitability the most; this can begin to lift farmers and rural households out of poverty, but on average it becomes more difficult to escape poverty (figure 1.2).

BOX 1.3   How many more are poor because of higher food prices?

Most analyses conclude that in the short term higher food prices raise the poverty headcount in most developing countries because not enough poor farming households benefit from the higher sales prices of their production (De Hoyos and Medvedev 2011, Ivanic and Martin 2008; Ivanic and Martin 2012a) to offset the negative impact of higher food prices on net consumers. This is so despite the well-known fact that three-quarters of the world's poor live in rural areas, and most of them depend on agriculture for their livelihoods.

A key to this apparent contradiction is that many of the poorest farming households are net buyers of staple foods. Over the long run, the negative impact of higher food prices on poverty is ameliorated through wage adjustments and household supply adjustments in response to rising food prices. Even in the long run, however, higher food prices appear to raise poverty in most poor countries and for the world as a whole—but the impact varies among population groups.

The two recent food price crises—in 2007–08 and in 2010—were researched extensively soon after they

occurred. The first crisis was estimated to keep or push 105 million people into poverty in low-income countries (Ivanic and Martin 2008), and the second crisis, 44 million people in low- and middle-income countries (Ivanic and Martin 2012a).

Using published information on the observed domestic price changes between June 2010 and March 2011[a] together with the techniques outlined in Ivanic and Martin 2012b, the implied poverty changes of the most recent food crisis was calculated (box table). The new estimates are calculated for the immediate short-run impacts, taking into account demand responses by consumers, medium-run impacts with wage adjustments, and long-run impacts including supply responses. The results suggest that changes in both wages and farmers' output responses reduce the negative impact of higher food prices on extreme poverty. But none of these long-run reductions is large enough to offset the initially large adverse impact on poverty.

Estimated poverty impacts of the 2010–11 food price crisis
Millions of people

Impact	2011 shock[a]
Short-run impact	48.6
Medium-run impact with wage adjustments only	45.5
Long-run impact including supply response	34.1

a. Refers to poverty change among low- and middle-income countries.

a. The calculations in Ivanic and Martin (2012a) used price changes until December 2010. The updated sample used includes more countries, 29, versus 9 than in 2008, and includes a range of household survey updates.

The aggregate impacts vary by region. Large net importers of food, such as those in the Middle East, North Africa, and West Africa, face higher import bills, reduced fiscal space, and greater transmission of world prices to local prices for imported goods such as rice and wheat. Higher prices particularly hurt consumers with high shares of household expenditure on food (as in many African and Asian countries). Large net-exporting countries, as in Latin America, Eastern Europe, and Central Asia, stand to benefit, in

part from potential higher tax revenues from (agricultural) commodities (figure 1.3 and box 1.5).

The pass-through of international to domestic prices has varied greatly across regions, with the largest pass-through observed in the countries of Latin America, which are largely open to international trade. In Sub-Saharan Africa the pass-through of rice and wheat prices to countries importing these cereals has been relatively fast. The transmission of international maize prices has

BOX 1.4   Actions by women made the most difference but were invisible to policy makers

Much of the response to a rise in food prices is reflected in additional care work by (mainly) women that is unpaid and not measured. Increases in food prices oblige women to invest greater time and energy to achieve the same level of nourishment and care of children, the sick, and the elderly. Examples of increased effort include more distant travel to hunt for bargains and more frequent shopping to purchase smaller quantities; more time devoted to chop or gather firewood because households can no longer afford other sources of energy; more time required to collect wild foods, and to beg and borrow money; having to undertake jobs considered hard or demeaning; and having to manage more stressful domestic family relationships, including drug or alcohol abuse, as well as violence.

One hopeful note is that across various community sites, parents and schools are working hard to keep children in school and provide essential food. Although teachers in Bangladesh and Zambia reported that local school dropout rates increased when food prices spiked, there was a much stronger emphasis than researchers expected on keeping children in school. In Kenya school feeding programs were often very accommodating of the poorest families, allowing them to bring younger siblings along at mealtimes. Nevertheless, higher food prices affect children's ability to learn. In Bekasi near Jakarta, for example, mothers were concerned that reducing pocket money for snacks was putting their children off going to school; some mothers in Kingston, Jamaica, had to pack children off to school with only a glass of water.

The food price crisis has meant that many poor people have suffered a serious decline in the quality and diversity of food, as well as in caloric intake. Food price increases directly reduced the quantity of food eaten by poor people and often forced them to eat food that was either unpalatable or unsafe. Some families made remarkable efforts to maintain nutritional levels: for example, in rural Zambia women replaced expensive small fish with protein-rich but cheap caterpillars.

As in previous crises, gender inequality can be expected to have increased in part because women have generally been the first to cut their food and other consumption in the face of falling real incomes. Nevertheless, in several communities a note of gender equality emerged, with young parents (particularly those not in manual jobs) stressing that both parents waited until their children had eaten well before themselves eating.

It is not surprising that the additional effort that (mainly) women have had to expend to cope with the food crisis has gone unnoticed. This phenomenon reflects a more general neglect of women's unpaid care work and the importance of its contribution to social protection and to the achievement of the Millennium Development Goals. A key lesson from this crisis should be that protecting progress toward the MDGs requires protecting caregiving. This should mean more direct support to women in their roles as unpaid caregivers, which entails recognizing and monitoring how their work is affected by food price volatility and other economic shocks.

Source: Oxfam, based on Hossain and Green 2011.

been much weaker, however, because most countries in Eastern and Southern Africa, main producers of white maize, fill their import needs through cross-border trade, not from overseas (Minot 2010). In Asia the transmission of changes of international rice prices to local prices differed significantly by country during the 2007–08 food price spike. In Bangladesh and Cambodia, the countries open to trade, the pass-through was fast and

relatively large, both immediately on the rise in the international price and three months afterward (table 1.2). The pass-through in China and India, the countries with high import protection, was small. Overall, for countries more open to trade (Dawe 2008; Robles 2011), and with a larger share of cereals imports in total domestic consumption, the faster and larger is the transmission of the international prices into local prices.

FIGURE 1.2   **The impact of higher food prices on poverty differs across socioeconomic groups**

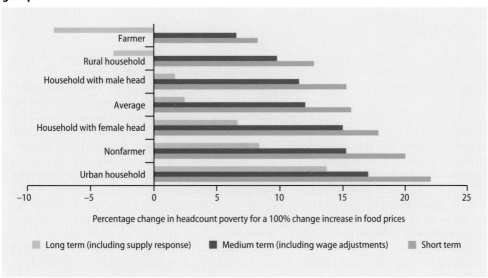

Percentage change in headcount poverty for a 100% change increase in food prices

■ Long term (including supply response)   ■ Medium term (including wage adjustments)   ■ Short term

Source: Ivanic and Martin 2012b forthcoming.

FIGURE 1.3   **Countries' vulnerability to global food price shocks tracked by share of cereal imports in domestic consumption and food share in household expenditure**

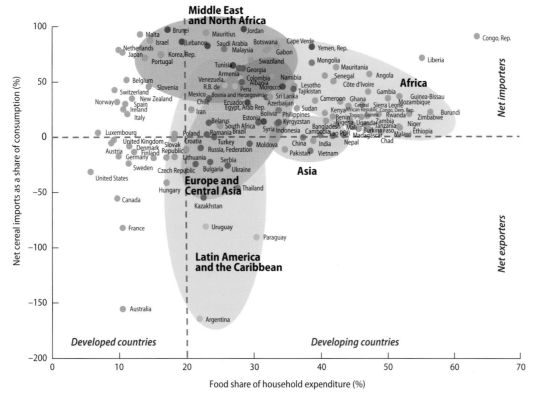

Source: World Bank 2011d.

BOX 1.5   World price impacts across regions

The spike in food prices in mid-2011 strained fiscal budgets, reduced incomes, and increased the vulnerability of the poor in many food-importing countries.

Sub-Saharan Africa. The region is particularly vulnerable to increases in international food prices, because in most countries some 50–70 percent of household spending is devoted to food, and the region imports about 45 percent of its consumption of rice and 85 percent of its consumption of wheat. High levels of malnutrition result in 38 percent of children being stunted. The situation is most perilous in the drought- and conflict-stricken countries of the Horn of Africa. Nevertheless, increases in cereal production driven by higher yields since the middle of the last decade improved the continent's ability to cope with the food price spike of 2011, compared with the experience in 2008. Governments should increase expenditures to raise the productivity of smallholder agriculture, strengthen trade between food deficit and surplus areas to reduce the volatility of local prices, and support the coping strategies of poor households in the face of continued food price volatility.

South Asia and East Asia and Pacific. South and East Asia are both self-sufficient in rice. Nevertheless,

some countries are net food importers, and the share of food in household expenditures remains about 40 percent in South Asia. Despite a mix of trade measures and buffer stock policies designed to slow the transmission of international to local prices (Dawe 2008), the 2008 food price spike significantly reduced household incomes in South Asia. At the same time, higher food prices increased fiscal deficits because of increased expenditures on food subsidy programs and safety nets (Ahmed and Jansen 2010). A dual approach of raising agricultural productivity and earned income, coupled with targeted safety nets, is needed to deal with hunger in South Asia. East Asia presents a different mix of challenges. Thailand and Vietnam provide over 50 percent of global rice exports and thus benefit significantly from rising prices; Indonesia and the Philippines are significant rice importers; and China is largely self-sufficient in rice. East Asia needs to maintain production while shifting to more environmentally sustainable processes in the face of increasing land and water scarcity (Christiaensen 2007).

Latin America and the Caribbean. Large resource endowments and the lower share of household expenditures devoted to food, at least compared with Asia and Africa, make the region as a whole less vulner-

TABLE 1.2   Pass-through of international rice prices to local prices in selected Asian countries
As a share of Thailand price (rice, 5% broken)

	Cambodia	Bangladesh	Philippines	India	China
Q2/07–Q2/08	98	55	63	25	23
Q2/07–Q3/08	79	60	46	37	25

Source: World Bank staff estimates based on FAO's Global Information and Early Warning System.
Note: The international price of rice (Thailand, 5% broken) peaked in April 2008.

High and volatile international food prices continue to be a big concern in the Middle East and North Africa, which is the largest wheat-importing region in the world. Some have even cited the food price developments since 2007 as a contributing factor in the Arab Spring (Breisinger, Ecker, and Al-Riffai

2011; Zurayk 2011). The long-term pass-through coefficients average 20–40 percent of the world food price increase, with the full transmission process taking about one year (World Bank 2011c). The pass-through effects are notably higher for West Bank and Gaza, Djibouti, the Arab Republic of Egypt, Iraq, and the United Arab Emirates. By contrast, in Algeria and Tunisia, the pass-through is small because of high food subsidies and controlled food prices.

Limited participation in international trade has led to higher local food price volatility, particularly in Sub-Saharan Africa. The price volatility of internationally tradable products is lower than that of nontradable commodities and commodities that are tradable

BOX 1.5   World price impacts across regions (continued)

able to volatile international food prices. However, agricultural production has been affected by natural disasters; for example, the January 2011 cold wave in Mexico damaged 1.5 million hectares (or 4 million metric tons) of white corn (for tortillas) and over 80 percent of green vegetable crops for export. And vulnerability differs significantly among countries. El Salvador, Grenada, Haiti, Suriname, and St. Vincent and the Grenadines are particularly vulnerable because of high fiscal deficits, large cereal imports, and low-quality social protection programs, while Argentina, Brazil, and Uruguay are agricultural powerhouses that benefit from higher international food prices. As a relatively urbanized region, a large majority of its population, including in net-exporting countries, are consumers who lose from the direct effects of price spikes (World Bank 2012a).

Europe and Central Asia. The region is quite diverse. Large grain imports and high shares of food in household budgets make Albania, the Kyrgyz Republic, Moldova, and Tajikistan vulnerable to rising food prices. By contrast, Kazakhstan, the Russian Federation, and Ukraine are food exporters that benefit from increased commodity prices. Similar to net-exporting countries in Latin America, net-exporting countries in this region with populations that spend significant shares of household budgets on food face

continued political pressure to impose export bans or to fix prices.

Middle East and North Africa. Countries in this region rely on food imports, particularly wheat, for at least 50 percent of domestic consumption. Thus, higher international prices can put considerable pressure on government and household budgets, depending on the level of domestic consumption subsidies and the pass-through from international prices. In the Arab Republic of Egypt, Djibouti, and the United Arab Emirates, more than 40 percent of a rise in international food prices is reflected quickly in domestic food prices, while in Jordan and the Republic of Yemen, countries with weak fiscal positions and a large dependence on food imports, the pass-through is 20–40 percent (World Bank 2011c). Higher domestic food production insulates Algeria and Tunisia from international price shocks. Oil exporters are well placed to cope with higher food prices, because their oil revenues have risen along with their food import bill. Since energy is an important input to agricultural production, increased oil prices have contributed to higher food prices.

Source: Updates by World Bank staff; World Bank 2011c.
Note: See the appendix for the current classification of economies.

only on regional markets. Wheat, rice, and cooking oil, products that are imported on the African continent, exhibit lower price volatility than the prices of domestically produced staples (table 1.3). The prices of maize, beans, and cowpeas, which are mainly traded locally and regionally, are more volatile, on average 20–30 percent above the price volatility of internationally traded commodities. Therefore, many African countries would benefit from reducing their protection levels and infrastructure costs to import from, or export to, international markets when needed to lower their high domestic volatility.

Higher food prices provide an opportunity for the private sector to produce and invest more and to improve productivity at

TABLE 1.3   Price volatility across products in the countries of Sub-Saharan Africa

Product	Number of observations	Number of price series	Volatility (%)
Tradable on international markets			
Wheat	224	3	9.4
Rice	2,202	30	10.8
Cooking oil	592	8	10.1
Nontradable or tradable only on regional markets			
Beans	878	12	13.3
Maize	3,450	47	14.4
Millet	2,224	30	10.5
Sorghum	1,914	26	12.4
Cowpeas	369	5	23.0

Source: Minot 2011, based on price data from the U.S. Agency for International Development's Famine Early Warning Systems Network.
Note: The local prices were analyzed from January 2005 to March 2011. Price volatility is defined as a standard deviation of logarithms of first price differences.

the same time. Higher food prices hurt poor net buyers, but increase agricultural incomes, and this in turn should provide incentives to expand production of the most profitable crops. Smallholder farmers in developing countries produce more when output prices improve (World Bank 2007). Higher staple crop prices in developing countries (25–35 percent higher in 2009 compared with 2006), and favorable weather contributed to higher production (5.2 percent), higher stocks (3.8 percent), and more trade (19.9 percent) in 2010–11 (FAO 2011a). High food prices offer opportunities for many poor countries to develop their agricultural sectors, which can help link local farmers to regional and global supply chains, increase local consumer access to competitively priced food products, and create new export sectors.

Agricultural productivity varies significantly across regions, indicating that improved use of existing technologies can lead to significant yield gains. For example, a comparison of current productivity with what is potentially achievable (demonstrated through on-farm research trials), assuming that inputs and management are optimized in relation to soil and water conditions, shows that the yield gap in maize production is greatest in Sub-Saharan Africa and lowest in East Asia. Yields in Sub-Saharan Africa are only 24 percent of what could be produced, while the gap is only 11 percent in East Asia (FAO 2011a). Better use of existing crop and nutrient management practices alone could increase rice yields in East Asian countries by at least 25 percent (Christiaensen 2011). About 15 percent of the value of the total rice crop in South East Asia could be saved through better post-harvest technology (especially drying and milling). A shift from area-based to volume-based charges for irrigation water in the Tarim Basin in China resulted in a 17 percent decrease in water use, while addressing poor land layout through adequate leveling and higher bunds to retain wet season water has been shown to increase yields in Cambodia by 27 percent.

Local conditions will determine the most effective mix of government policies in the face of food price spikes. In general, governments have a toolkit of various policy instruments to respond to food price spikes, and which combination to use depends critically on the initial conditions the country finds itself in, including its social and political environment. A major challenge has been to strike a balance between benefiting producers, and thus improving incentives for increased production, and protecting consumers within a macroeconomic framework that does not jeopardize fiscal and external sustainability.

Drivers of food price changes

Changes in world food prices reflect changes in food supply and demand and the corresponding responsiveness of the food system. World food price levels are driven over the long term by changes in demand from population and income growth, agricultural productivity outcomes, and secular changes in the prices of inputs, complements, and substitutes. Short-term shocks such as droughts, floods, changes in trade restrictions, volatile demand for associated inputs and outputs (such as oil and ethanol), and market expectations sharpened by low stock levels tend to drive food price volatility. The corresponding impacts on food prices are conditioned by the responsiveness of the food system, that is, the elasticities of supply and demand (table 1.4). The analysis focuses on cereals particularly because they are the most important staples for food security. The more responsive the system, the lower the corresponding impact on food price changes.

Longer-term trends in demand and supply

Recent growth of supply has outpaced growth in demand for main food crops (table 1.5). Increases in global demand for food are driven by population and income growth and by an accelerated use of food crops for industrial purposes, such as biofuels. Global food consumption growth over the past 50 years averaged 2.5 percent a year, or 1.4 times the average increase in population of 1.6 percent. Supply growth increased from an average 2.3 percent between 1960 and 2003 to

TABLE 1.4   Major drivers of world cereal prices

Average price levels		Price volatility	
Dependent on:		Dependent on:	
Long-term change in demand • Population • Income • Biofuels	*Long-term demand responsiveness/ elasticity to prices* • Share of food in consumption • Biofuels mandates • Oil/maize price ratio	*Short-term change in demand* • Oil prices volatility • Exchange rate volatility • Precautionary hoarding • Food reserves	*Short-term demand responsiveness/ elasticity to prices* • Stock release policies • Oil/maize price ratio
Long-term change in supply • Area planted • Yield changes	*Long-term supply responsiveness/ elasticity to prices* • Output and input market integration • Price risk management	*Short-term change in supply* • Droughts and floods • Share of production in more volatile production regions • Trade policy responses (export bans and sharp reductions in import tariffs)	*Short-term supply responsiveness/ elasticity to prices* • Trade openness

Source: World Bank 2012b, forthcoming.

TABLE 1.5   Higher consumption growth of corn has offset slowing growth in rice and wheat, while increases in area planted to food offset slowing yield growth

	Growth rate (%)					
	Demand			Supply		
Crop	1960–2011	1960–2003	2003–11	1960–2011	1960–2003	2003–11
Total (rice, corn, wheat)	2.5	2.5	2.5	2.4	2.3	2.8
Area				0.5	0.4	1.1
Yield				1.9	1.9	1.7
Rice	2.1	2.3	1.3	2.2	2.3	2.0
Area				0.6	0.5	0.9
Yield				1.7	1.7	1.1
Corn	3.0	2.8	3.7	2.9	2.7	3.8
Feed, residual	2.7	2.9	1.4			
Food, seed, industrial, including biofuels	3.4	2.8	7.7			
Area				0.9	0.8	1.8
Yield				1.9	1.9	2.0
Wheat	2.1	2.2	1.9	2.1	2.0	2.2
Area				0.2	0.1	0.8
Yield				1.9	1.9	1.4
Population growth	1.6	1.7	1.2			
Per capita income growth	1.4	1.4	1.5			

Source: U.S. Department of Agriculture; World Development Indicators database.

an average of 2.8 percent for 2003–11. More rapid growth in food demand than in population reflects higher demand for grain as animal feed (rising incomes increases the demand for meat) and the use of agricultural commodities in the production of biofuels. Future increases in demand will depend on changes in these three areas—food, feed, and industrial uses (biofuels). Population growth is now slowing, but demand for biofuels is rising.

There are significant differences in population growth across the globe. Sub-Saharan

Africa has the highest population growth (2.5 percent growth during the last decade), and Europe and Central Asia has the lowest (a mere 0.2 percent growth during the same period). Even though population growth might ease in Sub-Saharan Africa, current levels of population growth point to the need for this region, with its fragmented trade, weak infrastructure, low yields, and underdeveloped social safety nets, to address the bottlenecks of food security in an integrated but prioritized manner.

Increases in world food supplies depend on land area planted for food crops and subsequent yields. Average growth in grains supply over the past 50 years has been similar to growth in grain consumption (2.4 versus 2.5 percent a year, see table 1.5). Over this period 21 percent of the growth in grain production was from area expansion, while 79 percent was from yield improvements. However, during 2003–11, area expansion contributed 39 percent of supply growth while yield growth accounted for 61 percent; this shift is largely a reflection of a deceleration of yields and shifts of land away from the production of other crops to grains. Yield growth rates for rice and wheat have declined consistently with slowed development of higher yielding varieties and with an increase in production on more marginal land.

Land has become an increasingly limited resource, and the remaining arable land is almost by definition either less productive (inherently or requiring significant investment to raise yields) or, particularly in Africa, more difficult to exploit because it is located far from infrastructure. In the five years since 2005–06, land area for 13 major world crops increased by 27 million hectares, a rate that cannot be sustained indefinitely at the estimated supply of available land. Moreover, most of the expansion in land cultivation since 2005–06 (24 million of the 27 million increase) is located in only six countries or regions: China, Sub-Saharan Africa, former Soviet Union (Kazakhstan, the Russia Federation, and Ukraine), Argentina, India, and Brazil.

Future yield improvements may be harder to achieve than in the past. More binding land and water constraints, rising inputs costs, and lags in development of improved varieties may make yields gains harder to achieve. World yield growth rates have declined somewhat from 1.9 percent for the period 1960–2003 to 1.7 percent in 2003–11. Water constraints limit the future expansion of irrigated agriculture. Approximately 1.2 billion people live in river basins with absolute water scarcity, with the Middle East and North Africa and Asia facing the greatest water shortages and some greater scope for the expansion of irrigation in Africa.[3] Given continuing demographic pressures, particularly in Sub-Saharan Africa, it is important to increase land productivity, manage land sustainably, and improve the efficiency of water use. Rising populations mean that increasing food security to achieve reductions in poverty (MDG 1) may eventually conflict with ensuring the sustainability of development (MDG 7) (box 1.6) and the need for "green growth" (World Bank 2012d, forthcoming).

Short-term shocks in demand and supply

Food price uncertainty is rising. The uncertainty of food prices is driven by changes in both demand, including closer links to oil prices, exchange rate changes, and lower stock-to-use ratios, and supply, including weather, expansion of export crop production to areas where yields are less stable, switching of production to biofuels, and trade interventions affecting global supply.

Higher oil price volatility is contributing to higher food price volatility. The links between crude oil and agricultural markets have strengthened considerably since 2005, with the pass-through elasticity from crude oil prices to agricultural prices increasing from 0.22 in the pre-2005 period to 0.28 through 2009 (Baffes 2010). Crude oil prices increased sharply from early 2002 to mid-2008, more than doubling from early 2007. Crude oil prices have historically been more volatile than agricultural commodity prices, and the greater link between oil and agricultural markets will likely contribute to short-term food price volatility.

BOX 1.6   Sustainable increase in food production is required to simultaneously fight global hunger and reduce the pressure on biodiversity

The core of sustainable development is the challenge of fulfilling human needs and aspirations within the carrying capacity of our planet. This twin challenge is also reflected within the MDGs, with MDG 1—eradicating hunger—being part of a social floor, and MDG 7—ensuring environmental sustainability—addressing an environmental ceiling.

Failing to sufficiently increase production will have a backlash on the affordability of food and increase the risk of price volatility, thus reducing the stability of food supply. Agricultural area expansion to facilitate increased food production, together with other environmental pressures such as climate change and nitrogen deposition, results in further declining global biodiversity (PBL 2012). Having a long-term supply of food at reasonable and stable prices while at the same time halting global biodiversity loss requires that anthropogenic pressures on the environment be reduced. Measures include more efficient and better ecologically integrated farming, mitigation of climate change, improved land management, altered consumption habits—specifically a transition to low-meat diets in western countries—and reduced losses in the production chain, while increasing agricultural productivity (PBL 2010, 2012).

Two stylized strategies for increasing agricultural productivity within ecological limits could be followed: sharing or sparing. The first strategy focuses on mixing natural elements in existing and new agricultural areas and making optimal use of ecosystem services in agricultural production. Biodiversity impacts of expanded agricultural areas will be mitigated and reduced in existing agricultural areas, for example through edge effects and reduced fragmentation. The second strategy focuses on intensifying agricultural production in highly productive, already existing agricultural areas. Land conversion will be avoided as much as possible, settling for accelerated biodiversity loss in current agricultural areas while keeping a larger area of the world in its natural state.

The Netherlands environmental assessment agency (PBL) projects the required yield increase to address the twin challenge of eradicating global hunger while halting global biodiversity loss, by adopting a pure sharing or sparing strategy. The two strategies show different spatial patterns of biodiversity loss in 2050. For the sparing strategy, strict protection of natural areas is needed alongside a major effort to increase yields by approximately 1.3 percent a year globally. For the sharing strategy, intensive management of ecosystem services and landscapes is required, alongside investment in knowledge and practices on sustainable farming, to increase yields in a sustainable way by approximately 1.1 percent a year globally. These yield increases are comparable to those achieved in the 1970s and late 1990s but have to be maintained for a longer time period and in areas that did not have significant yield increases in the past.

Sustainable increase in food production to fight global hunger requires many simultaneous interventions, emphasizing the need for policy coherence. Creation of enabling conditions, knowledge transfer, and planning in accordance with physical potential are key. Areas where land management can be improved are tenure rights, regulatory institutions, and integrated planning. Furthermore, acknowledging the value and contributions of ecosystems and their services, especially to the livelihoods of poor people, is important.

Source: PBL 2012.

Declines in global stock-to-use ratios may be contributing to higher volatility. Historical evidence suggests that the likelihood of grain price spikes is higher when global stock-to-use ratios, a measure of physical liquidity of grain markets,[4] decline to low levels (Wright 2009; Stigler and Prakash 2011). Weather-related production disruptions reduced cereal stocks in developed countries by an estimated 28 percent between 2009–10 and 2010–11, in contrast to a 4 percent increase in stocks in developing countries. According to the Food and Agriculture Organization (FAO 2011a), the stocks of major grain exporters in 2011–12 are projected to decline further, reducing the global stocks-to-use ratio by 2.2 percent compared with 2010–11. Added to this is global uncertainty on the exact size and quality of stocks, uncertainty on the triggers for their release or buildup, and measurement revisions that can have significant market impacts. These concerns are particularly relevant for those countries that are highly dependent on food grain imports, as in the Middle East and North Africa region.

Adverse weather has played a significant role in the recent price spikes. Weather was an important factor in reducing production and stocks in 2010. The number of reported droughts, floods, and extreme temperatures seems to be increasing; in 2010 alone, a record number of 19 nations set temperature records. Recent extreme weather events include the Russian heat wave, dry weather in Brazil, and flooding in Australia, Pakistan, and West Africa. Floods are especially damaging, as they often require large reconstructions of irrigation systems and other infrastructure, and their frequency has increased along with the number of droughts. Overall, weather variability, possibly resulting from climate change, is having a significant impact on international food prices.

A larger share of world exports is being produced in more variable growing conditions. The major expansion of world grain exports in the last twenty years is in large part due to rapid increases in production for

export in the Southern Cone of Latin America. More recently, world markets have also become more dependent on supplies from the Black Sea region (Kazakhstan, Russia, and Ukraine).[5] The share of the Black Sea region and Latin America in global wheat exports doubled from 14 percent in 1990–95 to 28 percent in 2006–10. For maize, the share more than tripled, from 9 percent to 29 percent, over the same period. Yields in these newer export regions are less stable and overall supply and exports more variable, in part because of the willingness to use trade restrictions to ensure domestic supply, than in the traditional breadbasket areas of the developed world. Thus global supply of these crops is likely to become more variable over time, contributing to potentially higher volatility in world food export volumes and prices.

Insulating policies reduce the role that trade between nations can play in bringing stability to the world's food markets.[6] Open trade policies are essential to provide positive incentives to national producers of food and to attract investment from all sources. Although exporters and importers have possibly been more restrained than in 2008, insulating trade interventions was nevertheless still widespread and even rose in 2011 (versus 2010), continuing to contribute to price instability.

The inelastic nature of world food demand and supply lead to large price increases from shocks to the system (that it, the system has limited flexibility to respond, at least in the short term). Over time, world food demand will become more price inelastic as incomes rise, and, if not offset by a more elastic supply response, price increases per demand and supply shock will be higher in the future than in the past.

World price elasticities of food demand are low and tend to decline with increases in per capita income (figure 1.4). The increased demand for biofuels can influence this long-term trend in two ways. First, biofuels mandates act to fix demand for corn-based ethanol (at any price), thereby further reducing overall demand responsiveness to price changes. Second, if long-term oil prices

FIGURE 1.4   Demand responsiveness to food price declines as per capita income increases

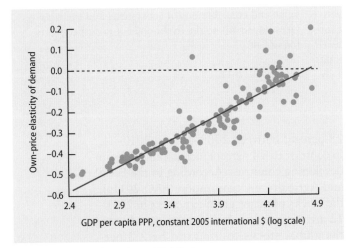

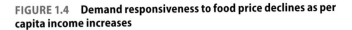

Source: U.S. Department of Agriculture and World Bank.

rise dramatically, making corn-based etha-nol profitable beyond the mandates, then the overall demand responsiveness to price changes could increase (oil prices relative to corn have been higher). The net effect on price responsiveness depends on which of these two effects dominates.

Long-term supply responsiveness to price changes is influenced by output and input market integration and price volatility impacts on production decisions. The world food-supply response is estimated to be low (with estimated price elasticities of 0.1 per-cent). Price elasticities tend to be higher in developed than in developing countries, in part because of more developed and inte-grated input and output markets. In addi-tion, higher price volatility in food markets increases risk and likely lowers the produc-tion response to higher prices (as it does for other crops in developing countries; see Subervie 2008). While the longer-term supply response may rise as countries develop (with greater output and input market integration), this may be offset by lower supply response induced by higher price volatility (and more constrained land).

Recent policy responses

Some responses taken by developing-coun-try governments have not been conducive to longer-term growth. An FAO review of policy responses by 81 developing countries to the 2006–08 price spike showed that a large majority of countries used distortion-ary measures that could undermine agricul-tural productivity over the long term (table 1.6). Nearly 70 percent of countries used trade policy instruments, such as reductions in import tariffs and export taxes or bans, to reduce domestic prices. Many combined trade policy instruments with domestic measures, such as reductions in food taxes, release of stocks at subsidized prices, and price administration, to lower food prices for all consumers at the expense of producers. Half of the country sample used safety nets to mitigate the impact of rising food prices on the most poor and vulnerable, while allow-ing domestic prices to rise to induce a food supply response. Where local prices were reduced, governments often provided support to producers in compensation for the lower output price, but production support rarely

TABLE 1.6   Policy measures adopted in 81 selected countries in response to 2006–08 price spike

Policy measures	Regions (number of countries surveyed)			
	Africa (33)	Asia (26)	Latin America (22)	Overall (81)
Trade policy				
Reductions of tariffs and customs fees on imports	18	13	12	43
Restricted or banned export	8	13	4	25
Domestic market measures				
Suspension/reduction of value added or other taxes	14	5	4	23
Released stocks at subsidized prices	13	15	7	35
Administered prices	10	6	5	21
Production support				
Production support	12	11	12	35
Production safety nets	6	4	5	15
Fertilizers and seeds programs	4	2	3	9
Market interventions	4	9	2	15
Consumer safety nets				
Cash transfers	6	8	8	23
Food assistance	5	9	5	19
Increase of disposal income	4	8	4	16

Source: Demeke, Pangrazio, and Maetz 2009.

TABLE 1.7   Fiscal implications of policy responses to 2006–08 price spike, selected countries

Fiscal costs	Year	Argentina	Brazil	Chile	China	India	Indonesia	Russian Federation	South Africa	Ukraine	Vietnam
Total fiscal costs	2007	49	743	0	436	5,273	644	−32	786	79	48
(US$, millions)	2008	−122	2,394	56	7,813	24,000	2,095	2,309	1,849	246	242
Share of fiscal	2007	0.1	0.2	0.0	0.1	3.8	0.8	0.0	0.9	0.2	0.3
revenue (%)	2008	−0.1	0.6	0.1	1.7	19.1	2.1	0.6	2.4	0.6	1.0
Fiscal cost per person	2007	3	5	0	1	12	6	0	27	4	2
(international $, PPP)	2008	−5	16	5	11	55	16	22	67	10	7

Source: Jones and Kwiecinski 2010.
Note: PPP = purchasing power parity.

has been large enough to fully make up the loss from lower output prices.

The fiscal costs of policy responses have varied, depending on the mix of instruments used. Brazil and Chile, for example, focused on safety nets to protect vulnerable consumers. The additional fiscal cost involved was 0.1 percent of total fiscal revenue in Chile and 0.6 percent in Brazil in 2008 (table 1.7). In South Africa, which followed similar policies, the fiscal bill of 2.4 percent of revenues was larger because of smaller total fiscal revenues and the larger number of beneficiaries. India, which provided short-term stimulus to food and fuel price spikes, incurred the largest fiscal response cost. In most other emerging economies, the fiscal cost of response was about 0.5 percent of total budget revenues.

Sustainable policy responses

The most appropriate policy mix to address the implications of higher and more volatile food prices is determined by the characteristics of each individual country. The chosen policy mix at the country level depends critically on how much of a country's food needs to be imported, how much of their income the poor spend on food, and the socioeconomic characteristics of the poor affected. It depends equally on a country's integration with regional and world markets, its level of productivity compared with what is achievable, and its government's capacity to target the poor and vulnerable through mitigating interventions. In addition, much is contingent

on a county's initial macroeconomic condition and thus its ability to expand public expenditure programs or provide tax incentives without jeopardizing fiscal sustainability. Hence, the content of the policy interventions chosen will be roughly the same from country to country, but the sequencing and priority given to each intervention and its magnitude will differ.

A policy mix should contain a combination of short-term measures to alleviate the immediate hardship on the poor and vulnerable and long-term measures that address the main bottlenecks to the functioning of the domestic food markets and profitability of farmers. In the short term, much depends on the ability to alleviate immediate poverty implications of higher food prices through social safety nets and efforts to increase agricultural production quickly. Over the long term, policies should focus on limiting the average rise in food prices and food price volatility.

Measures to reduce the negative impacts on food security in the short term

Governments will need to consider the different implications for the various socioeconomic groups when designing an effective policy response. While various policy actions can be instigated to prevent future food price spikes, measures to mitigate the immediate adverse impacts can and should be taken to protect the poor and vulnerable. However, various socioeconomic groups are affected differently, and any policy actions taken should be informed to the extent possible by information about the groups most affected. Even though lower taxes can lower food costs to consumers, they are often not well targeted, and consequently large amounts of scare public resources flow to higher-income consumers. In addition, the choice of actions in the short term should not undermine longer-term farm incentives to invest and produce more; both export bans and ad hoc provision of inputs can have deleterious effects on farm incentives.

The urban poor are usually net consumers of food and have little opportunity to increase subsistence food production. Hence, assisting the urban poor depends almost entirely on social safety net programs. However, national programs are often oriented to rural areas, where the share of the poor in the population is commonly higher (Baker 2008). At the same time, the urban poor often live in informal settlements and are more transient than the rural population, and therefore expected economies of scale with urban social safety net programs are not always realized.

Female-headed households are more vulnerable. Women are in general more vulnerable to economic shocks, and various gender-based vulnerabilities, including extensive time burdens, limited legal benefits and protection, and limited access to financial resources (World Bank 2011e), make female-headed households even more vulnerable. Policies to help female-headed households and women in general weather a food price crisis must be tailored to the specific socioeconomic and cultural context in which gender relations unfold. For example, food-for-work programs could scale up lighter tasks suitable for women, and conditional cash transfers could provide higher benefits to girls, who are more likely to be kept out of schools.

Social safety nets have a vital role to play in coping with food price shocks. Social safety nets can be used to protect the poor by providing conditional or unconditional cash transfers, offering short-term employment, and discouraging negative mechanisms for coping with the setbacks caused by a food price crisis. Where markets are functioning well, cash may be more effective than food in providing assistance but may leave poor people exposed to price risks. When food markets are functioning poorly, or where prices are increasing rapidly, food transfers may be a more effective means of assisting the poor and vulnerable (WFP 2008). Cash or food-for-work programs that develop infrastructure should consider implications for future maintenance, and opportunities to develop skills in the types of work selected (such as road paving). Countries that had prepared permanent social safety programs and institutions during good times were better positioned to scale up as needed than those that had not (box 1.7). Thus middle-income countries, where social safety net programs are relatively common, were often better placed to support the poor during a food price shock than low-income countries (although such programs are increasingly being adopted by more low-income countries).

Using an effective social safety net program to address a food price crisis depends on the programs that already exist and the capacity on the ground. Establishing a single social safety net program may be a priority in a low-capacity setting, while in a middle-income country the priority may be ensuring that different programs coordinate well with each other and target the identified and intended beneficiaries, particularly female-headed households and the urban poor (box 1.8 and World Bank 2011b). The various food, fuel, and financial crises over the past decade have underlined remaining weaknesses in the effectiveness of social safety nets, which differ by countries' income level. Middle-income countries had trouble with increasing coverage or benefits as needed, while low-income countries often lacked poverty data and systems to inform the choice of a particular social safety net program and of ways to target and deliver benefits. Programs that deal with chronic poverty are not

BOX 1.7 Ethiopia's food security programs

Home to 79.1 million people, Ethiopia has achieved steady, two-digit economic growth in the past few years, lifting many out of poverty. However, with a growing population, inadequate infrastructure, low agricultural productivity and recurring droughts, floods, and land degradation, 15 million people remain poor and vulnerable to food insecurity. In the past two decades, emergency food aid dominated responses to food insecurity in Ethiopia. Yet such aid was often unreliable, arrived late at a daunting cost, and focused on immediate relief assistance at the cost of improving overall livelihoods. In response to the growing consensus on the need for reform, the government decided to launch the Food Security Program in 2003, composed in part of a Productive Safety Net Program (PSNP) and a Household Asset Building Program (HABP).

The Productive Safety Net Program (PSNP)

The PSNP aims to improve food security by providing short-term transfers that help prevent asset depletion at the household level and by creating assets at the community level to ensure against unpredictable shocks. The program consists of two components: direct transfer support and labor-based public works. The direct support program provides predictable and timely cash and food transfers to chronically food-insecure households and extends the option of participating in community work (in child-care centers and nutrition education). The public works program is focused on creating sustainable community assets, mainly aimed at rehabilitating environmentally degraded areas and developing watersheds, with the core objective of increasing productivity and providing sustainable livelihoods.

The PSNP was launched by integrating existing government agencies and entrusting them with program implementation. Importance was placed on capacity-building initiatives within the agencies, in addition to creating horizontal links to avoid forming parallel structures. At the same time, 10 donor organizations agreed on a harmonized government-engagement model, by forming a joint coordination committee to oversee the programs' implementation.

Soon after, the donors adopted both financial management and procurement frameworks, on top of the single monitoring system, to ensure that the programs are kept on track. To date, with the help of the donor organizations, the PSNP covers 8.3 million people across 318 districts.

Achievement and lessons. Since the beginning, the PSNP has proved to be instrumental in supporting beneficiary consumption, protecting household assets, and building community resources. From 2005 to 2009, PSNP interventions enabled 75 percent of targeted households to consume more or better quality food; 62 percent to avoid selling assets; 23 percent to acquire new assets; 46 percent to use more health care; and 39 percent to send more children to school. Overall, the PSNP experience proves that a safety net program in a low-income setting can be implemented by multiple organizations and have multiple funding streams. The PSNP also demonstrated that predictable cash transfers are key determinants of a cash transfer program's impact; that the sustainability of public works programs depends on local management; and, above all, that there is political will and capacity to move away from one-time humanitarian response programs, to more sustainable development-oriented interventions.

The Household Asset Building Program (HABP)

The HABP aims to help households graduate from the PNSP and to assist recent graduates. Within the HABP, a household is considered to have graduated when it becomes food sufficient; that is, when, in the absence of receiving transfers, the household is able to meet its food needs for 12 months and withstand modest shocks. Overall, the HABP program seeks to diversify income sources and increase productive assets of food-insecure households that are, or have been, PSNP beneficiaries. It focuses on facilitating the beneficiary households' access to on- and off-farm inputs, technology, and financial services in order to graduate from the program.

Source: World Bank country teams.

BOX 1.8   Building foundations for social safety net systems

While many of the initial experiences with safety nets involved ad hoc responses to crisis, it has become clear that building effective safety nets within a broader social protection system is essential. Critical building blocks for an effective system include:

- Identification: Mechanisms to identify eligible beneficiaries and promote empowerment should be established.
- Targeting and eligibility: Simplified approaches drawing on available information, bearing in mind costs, should be used.
- Enrollment: Either a census-style survey or an on-demand system may be used effectively. Each can be appropriate at different stages of program development, or they can be used simultaneously.
- Timely payments: New technologies can help, but simple, traditional systems can also work.
- Monitoring and evaluation: Basic monitoring systems should be established, as a foundation for immediate impact evaluation and to establish the database required for future evaluation.

Financial sustainability is a key issue, because programs usually have external financing for only a short period of time without a guaranteed government budget for the longer term. The high level of fragmentation of sources of financing and programs make planning and budgeting more complicated, and hinders domestic ownership of social protection programs. Donors have attempted to address these challenges through new aid modalities that move away from fragmented project aid toward general budget support and sectorwide approaches. However, examples of

successful coordination and pooling of resources are not very common, especially in low-income settings.

Successful coordination depends on countries themselves taking the lead in creating joint processes for developing strategies and programs, as well as encouraging donors to harmonize their policies. Sound public financial management systems, periodic reviews with performance indicators, independent procurement audits, targeting and process evaluations, appeal mechanisms, community monitoring, and perception surveys are all tools that can be used to strengthen mutual confidence between government and donors.

Credible and transparent systems help ensure program effectiveness and sustainability. The rapid expansion of safety nets has spurred the need to ensure efficient and effective use of public funds. It is important to define clear roles for each institution (including public, private, and donors) in coordination and execution of social protection reform, taking into consideration capacity levels and political weight. Critical ingredients for promoting transparency and accountability are:

- Strong controls: Accountability measures are required from top down and bottom up.
- Clear roles: All actors should understand how they fit into the system and their responsibilities. Local community, private organizations, and social funds can all be used to enhance strong governance.
- Well-communicated rules: Clear operational guidelines should be disseminated to all actors.

Source: World Bank 2011a.

necessarily well suited to address the transient impact of shocks on the poor (Alderman and Haque 2006). Existing social safety nets provide a basis to scale up implementation and coverage in the event of excess need. Relatively small-scale programs may provide the administrative infrastructure, including the rules of operation and eligibility, that can

be adapted to a major crisis without costly implementation bottlenecks.

Actions to increase the agricultural supply response in the upcoming season can help reduce interseasonal impacts of price spikes on food security. Targeted input support can enhance the ability of smallholders to respond. Provision of inputs works best when it mobilizes the private sector (through vouchers, for example) and is complemented by reductions in logistical overheads, especially in ports and on roads. Anticipating and enlisting policy support for dealing with potential bottlenecks that restrict delivery of inputs to national borders are essential. In addition, demand estimates for fertilizer and seeds need to be periodically reviewed in an environment of rapidly changing inputs prices to prevent waste from overestimates and constrained impacts from underestimates. Meeting these requirements is key to generating value for money from public expenditures on inputs.

Agriculture can contribute to gender equality by improving access to economic opportunities for women, which also would increase agriculture productivity. Women now represent 40 percent of the global labor force and 43 percent of the world's agricultural labor force. Productivity will be raised if their skills and talents are used more fully and through projects that are gender sensitive in both design and implementation. The FAO estimates that equalizing access to productive resources between female and male farmers could increase agricultural output in developing countries by as much as 2.5 to 4 percent. For example, if women farmers were to have the same access as men to fertilizers and other inputs, maize yields would increase

by almost one-sixth in Malawi and Ghana (World Bank 2011e).

Measures to address the drivers of higher and more volatile world food prices in the long term

Demand- and supply-side responses can help to reduce future food price escalation. Responses are needed today at global, regional, and local levels to have the anticipated impact in the long term. While the global supply of cereals has outgrown aggregate demand during the past eight years (see table 1.5), and while a few of the large and technology-intensive exporters, such as the United States, retain significant capacity to expand production in the near- to mid-term, ensuring sustainable supplies of food at the local level requires improvements in agricultural productivity and facilitation of trade in and among developing and developed countries (table 1.8). Measures include promoting increased yields through research, extension, and improved water management; improving the efficiency of land markets and strengthening property rights; addressing biofuel mandates and improving cost-efficiency of biofuels technologies; increasing farmers' access to efficient tools to manage risk; and increasing the integration of domestic with world markets. Policies that would limit

TABLE 1.8   Main measures to limit the growth and volatility of world cereal prices

Measures to reduce price volatility		Measures to reduce average price escalation	
Short-term changes in supply • Development of more weather-tolerant varieties	*Short-term change in supply responsiveness to prices* • Trade openness	*Long-term change in supply* • Raised crop yields • Improved water management • Improved (rural) investment climate including through: – Improving access to finance – Facilitating land markets	*Long-term supply responsiveness to prices* • Better use of price risk management tools • Strengthened market integration, including infrastructure and private-sector development
Short-term changes in demand • Increased transparency of agricultural markets	*Short-term demand responsiveness to prices* • Efficient food reserve management	*Long-term change in demand* • Shifts to market-based biofuels policies and promotion of more efficient technologies	*Long-term demand responsiveness to prices* • Shifts to market-based biofuels policies and promotion of more efficient technologies

Source: World Bank 2012b, forthcoming.

food price volatility include the development of weather-tolerant price varieties, increasing the size and improving the management of stocks, opening markets to trade, and improving market transparency.

Several actions can directly address volatility

Public investment to develop more weather-tolerant varieties can be increased. Weather-tolerant crop varieties can reduce food production and price shocks. Many studies have found that use of drought-resistant maize varieties can increase yield by as much as 40 percent under drought conditions in Sub-Saharan Africa. Similarly, breeding millet and sorghum for drought resistance has produced yield improvements of as much as 50 percent. Substantial room also remains for research on transgenic methods to improve the drought resistance of crops in semi-arid regions. Transgenic drought-resistant maize varieties are found to yield up to 20 percent more than nontransgenic drought-resistant varieties (Kostandini, Mills, and Mykerezi 2011). Transgenics is indeed an underutilized technology for poverty reduction. Because of the potential risks involved, however, it should be implemented only in situations where international biosafety standards are in place.

Public food grain stocks can be used effectively to reduce domestic and world food price volatility. Sufficient stock levels can reduce the likelihood of price spikes, and good management, particularly of purchases and releases, can reduce rather than amplify volatility. But stocks always cost money, which can be as high as 15–20 percent annually of the stocks. Costs are high, while benefits in terms of price stability and economic growth are realized only when stocks are well managed (World Bank, 2012c, forthcoming). Further technical and consistent guidance to national governments on costs and benefits, levels, and use of food stocks is needed.

Small emergency public food grain reserves, at the national and regional levels, related to the consumption needs of the most vulnerable, have an important role to play in alleviating the consequences of high and volatile prices, if they are well targeted to the most vulnerable people. In contrast, using stocks as an instrument of domestic price stabilization has proven difficult because of their high costs—both financial costs (implicit interest, hidden quality losses, physical storage losses, and transaction costs of stock rotation) and efficiency costs through disincentives to (generally more efficient) private-sector storage and trade (Dorosh 2009).

Open trade across all markets can diversify short-term production shocks, thus dissipating the associated price effects. Price insulation reduces the ability of world markets to dissipate shocks, and trade barriers imposed in 2007–08 acted to amplify the food price spike rather than reduce it. Trade is even more important when food stocks are low, because more countries need to enter markets as net buyers. Improved social protection policies in net-food-exporting countries (particularly for large exporters like Argentina, Kazakhstan, Russia, and Ukraine) would reduce pressures for export restrictions when food prices rise. Continued analysis of the likely gainers and losers from trade policy changes would help guide government policies and trade negotiations.

Greater market transparency would reduce market uncertainty and the associated large price corrections following revisions to market information (production, stocks, and trade). Clearer and more accurate monitoring can help to reduce food price spikes. The capacity of international and national providers of food market information, public as well as private, to monitor market developments and disseminate timely and accurate information on food prices and food security should be strengthened.[7] A good step in this direction is the establishment of the Agricultural Market Information System (AMIS).[8] AMIS is a major partnership effort of multilateral international organizations to leverage their scarce resources and to use the comparative advantage and expertise of different organizations to improve global short-term agricultural forecasts and policy analyses of global production, trade, stocks, and price developments; and to promote early information exchange and discussion on crisis prevention and responses among policy makers through a Rapid Response Forum. More efforts are needed to ensure that better market information is shared and used for agricultural policy decisions. Initial commodities to be tracked are wheat, rice, maize, and soybeans.

Measures to reduce average world food price escalation

A broad range of actions is needed across both developed and developing countries to reduce pressures on food prices. Developed-country policy reforms would likely reduce average world food price increases (with higher world food prices from tariff and subsidy reforms (World Bank 2007) being offset by lower prices from biofuel policy reforms). Middle- and low-income countries can play a significant role in the supply response, enhanced by improved policies and investment in productivity growth. Middle-income countries including Argentina, Brazil, Uruguay, Kazakhstan, Russia, and Ukraine have significant potential for productivity gains and have accounted for a larger share of recent global food exports. With macroeconomic stability, lower conflict, and lower agricultural taxation, agricultural growth itself and its potential growth in Africa is also improving. But more is needed, particularly through more and better public and private investments.

Closing the gap between average farm and experimental food crop yields can greatly contribute to a solution to regional and global food security. More and better public and private investments are needed to increase adoption of improved technology, to generate new and improved technologies, to improve agricultural water management and the efficiency of irrigated areas, and to increase economies of scale in farm production and processing though private-public partnerships. This agenda is particularly relevant for countries and regions, such as Sub-Saharan Africa where yield gaps are large and adaptation of new technologies

has been lagging. An important part of this agenda includes adaptation of high-yielding varieties with resistance to biotic (pest and disease) and abiotic (climate change) stresses; improved soil fertility through crop rotations and judicious use of organic and inorganic fertilizer; and better integrated management of pests, diseases, and weeds in conjunction with more efficient water management (FAO 2011c). Complementary investments will be needed to better align extension services with farmers' needs, supplemented with better use of information and communication technologies, increased use of matching grants for technology adoption, and strengthened seeds and fertilizer markets.

Investments in improved and sustained water management can enhance the returns to investments in other soil and crop management practices. Greater attention is needed to ensure sustainable water management practices through water use associations; incorporation of broader river basin management aspects; and improved use of shared watercourses, including support for cooperation between different riparian states on the use of scarce resources. Expanded irrigated areas and improved water use efficiency of existing schemes are both needed, as is better water control and erosion prevention at both field and river basin levels. In Africa, a lower share of cultivated land is irrigated, leaving its food system more vulnerable to climate risks. With climate risks expected to increase, it is important to take advantage of higher food prices and thus improved profitability of irrigated agriculture, to attain better water management in food production through investment in irrigation, and thus the higher productivity and reduced variability that irrigated production systems enable.

Public actions to induce a private-sector-led supply response may need improvement in the investment climate. To orchestrate a supply response, each country will need to ensure that the private sector can take advantage of the higher prices. Issues that often affect a (rural) investment climate include access to finance, (land) property rights, various licensing and registration requirements, sector specific regulations, and taxes and tax administration. Addressing these potential bottlenecks will reduce the cost of doing business and increase competition.

Access to finance can greatly improve farmers' ability to take advantage of higher prices and improvements in the country's economic policy environment and economic infrastructure. However, because most rural households lack access to reliable and affordable finance for agriculture, the improved economic environment does not automatically translate into higher private investment. Many small farmers live in remote areas where retail banking is limited and production risks are high. The recent financial crisis has made the provision of credit even tighter and the need to explore innovative approaches to rural and agricultural finance even more urgent.

Facilitating land markets can expand the areas sown to food crops and improve yields. Land sales, more efficient rental markets, and strengthened property rights can improve the productive efficiency of existing land areas and make better use of remaining areas available for crop production. Secure property rights are also a prerequisite for land consolidation where it is needed. Attention is needed to ensure responsible agro-investment from foreign investors and to secure the land rights of poor farmers. Increased foreign investments may spur agricultural productivity growth, fiscal revenue, employment, and local incomes, but may also result in local people losing land on which their livelihoods depend (Deininger et al. 2011). Capacity strengthening is needed to ensure that the terms and conditions of land deals enable local (farming) communities to seize opportunities and mitigate risk.

Strengthening property rights, particularly for poor farmers can improve the use of existing cropped areas. Making land rights more transferable increases investment incentives and allows access to land through sales, rental markets, or public transfers. In some countries, particularly in Latin America and southern Africa, inequality in land ownership often leads to underuse and deep-rooted rural poverty. In such cases, increased access through targeted programs of financial assistance to enter land markets can potentially increase productivity and promote equality. Land programs also help agricultural regions to rebuild after conflicts and natural disasters, such as in Sri Lanka and Aceh, Indonesia. Significant gains can therefore be generated from land policy and legal reforms; increased security of existing customary or informal land tenure; modernized land administration; land redistribution through socially manageable processes; and prevention and reduction of land conflicts through dispute resolution mechanisms among other means.

Reducing biofuels mandates and promoting more efficient technologies can reduce escalation in food demand for industrial purposes. The six largest producers account for about 95 percent of world biofuels production. In 2010–11 an estimated 37 percent of all maize used in the United States, the largest user of maize for biofuels, went into making ethanol (Trostle et al. 2011).[9] Policies to promote biofuels have included crop production subsidies, infrastructure for biofuels storage, blending and production mandates, import duties, and tax incentives. These policies have provided overall support for ethanol worth $0.28 a liter in the United States and $0.60 a liter in Switzerland, and for biodiesel, $0.20 a liter in Canada and $1.00 a liter in Switzerland (Steenblik 2007).[10] While biofuels offer a source of renewable energy and possible large new markets for agricultural producers, current biofuels programs have a mixed record of financial viability without subsidies.[11] Because ethanol demand and corresponding prices have been raised by government regulation, deregulation is part of the solution to reducing food price escalation. Removing both nonmarket actions to raise demand for biofuels and subsidies for its production can reduce competition for grains among fuel, food, and feed. Open international markets should be encouraged so that production of biofuels occurs where it is economically, environmentally, and socially sustainable to do so (G-20 2011). At the same time, countries should focus on generating new technologies that need fewer agricultural commodities to produce biofuels.

Ensuring a food supply response to higher prices, and greater participation of smallholder farmers in this supply, requires better use of price risk management tools to reduce uncertainty. Earlier analysis showed that developing-country crop supply response declined significantly when price instability doubled, but that use of risk management tools (such as precautionary savings and access to financial services) reduced the negative impact of price volatility on production decisions. Improved farmer access to price risk management tools can help ensure supply response to higher prices (help prevent a decline in the price elasticity of supply) (box 1.9). Improving access of smallholder farmer and microenterprises to financial services for agriculture and food retail through direct service provision, market facilitation, and an improved enabling environment will likely have a broader impact than would improving access to more formal price-hedging instruments (such as commodity exchanges or warehouse receipts). Traders have typically used formal hedging instruments more than farmers, although basis risks (price correlation between domestic markets and the closest futures market) are often too high to justify their use. These risks can be lowered, but doing so often requires complementary long-term investment in transport infrastructure.

Better market integration ensures that world price signals reach more producers and thus induce a supply response, thereby increasing the responsiveness of the food system to price increases. By linking farmers more closely to consumers, marketing systems can transmit signals to farmers on new marketing opportunities and guide their production to meet consumers' preferences. Strengthening the links between local suppliers and food retailers can help to provide locally produced goods at more competitive prices. Consequently, public and private investments are needed to expand the reach and quality of rural roads, improve the collection and dissemination of market information, including through information and communication technologies, and improve technologies for post-harvest storage to reduce product losses. In addition, investing in agribusiness logistics and distribution infrastructure through private-public partnerships can facilitate trade, lower costs, and reduce post-harvest waste. Strengthening the bargaining power of smallholder farmers—especially women—through their producer organizations can help further reduce transaction costs, improve economies of scale, and hence better link them to markets.

These measures will help both small and large farms. However, the sector dominated by smallholders will require more public goods from the government than the sector dominated by larger farms. This is because the provision of agricultural services to small farmers presents significant coordination challenges and thus high transaction costs for the private sector. While large farms need a basic enabling environment to facilitate access to the most important production and marketing support services (capital, inputs, technical and market knowledge, marketing contacts) on their own, various public interventions are still required to ensure that these services are provided to smallholders, including through public-private partnerships. This task is more challenging but has high pay-offs.

Poverty implications of higher agricultural productivity in developing world

Agricultural prices and price volatility are likely to remain high. Official forecasts suggest that fundamental factors will keep global prices higher than pre-2007 levels over the medium term (G-20 2011; World Bank 2011b). Accelerated use of food crops for industrial purposes (biofuels) continues to offset the effect of slowing population growth on food demand. And production gains may be harder to achieve in the future, with more limited space for area expansion, declining yield growth, and increased weather variability. High price volatility will likely continue because world stocks remain low and the low responsiveness of the food system amplifies

BOX 1.9   Managing supply and price risks for maize in Malawi

High international and regional prices have created an export opportunity for Malawi. But these higher prices can also translate into higher risks if the country experiences grain shortages. One strategy to cope with these risks is to strengthen domestic market demand and stockholding with a repurchase option (REPO) deal. REPOs involve agreements between government and banks or grain traders for the bank or trader to purchase maize during the harvest season (June/July), hold stocks in the country, and later sell these stocks to the government at a pre-agreed price on a stipulated date in the future (such as January/February) if the grain is needed. If the grain is not needed, the bank or trader would expect to export it to neighboring countries.

The REPO contract offers Malawi several advantages that contribute to price and supply stabilization.

• The contract has a stipulated grain purchase price that can be used as a reference point for any purchases by Malawi's agricultural marketing agency. The additional demand for grain created by this deal would help support a floor price at harvest time.
• Malawi could more readily take advantage of regional grain demand by encouraging exports—with the knowledge that the country would maintain adequate stocks for its own requirements. This, again, would contribute to the strengthening of producer prices.
• The REPO would create a second layer of grain stocks in the country held in complement to the holdings of the Malawi's National Food Reserve Agency. Depending on how the deal was managed, this could encourage a broader range of traders to hold grain stocks in rural areas. The stipulations of the contract would ensure these stocks were maintained in good condition.
• Finally, the grain would be readily available in the country if the next cropping season started poorly. If stocks appear adequate in the country, and the next season starts well, this grain can then be exported.

As an example, in May 2007 a repurchase option (REPO) deal would have provided financing for a purchase of up to 150,000 metric tons of maize in July at a price of MK 18–19 a kilogram. The government would have had to pay a premium of MK 3.4 a kilogram for a bank or trader to hold these stocks in

Malawi for up to seven months—for example, until January 2008. The government would then have had the right to repurchase the maize at an agreed price of MK 25 a kilogram and to use this maize to resolve any localized supply shortages. Such as step would have helped limit any rise in retail grain prices during the lean season. It also would have contributed to reducing the price volatility seen that season (MK 14 a kilogram in July, but MK 35 in January and February). Alternatively, the government could have simply allowed the grain to be exported.

The REPO deal and similar supply/price management contracts fit into a toolbox of complementary risk management strategies designed to reduce price variability and strengthen domestic markets. Other tools include the following.

• *Weather insurance* can provide funding for imports in the event of severe production shortfalls associated with drought. Index-based weather insurance can be used to insure individual farmers, guaranteeing them an income in the event of a drought.
• A *warehouse receipts initiative* can improve the availability and quality of warehouse facilities for grain trade, reduce grain storage losses, and improve the availability of finance for the market.
• A strengthened *market information system* can improve price transparency and alert traders to opportunities for moving grain from surplus to deficit regions.

Ongoing Bank work has yielded valuable lessons about constraints to hedging food prices. Lessons include the following.

• Many governments are not focused on ex ante management of food price shocks and are not assessing the risk as a contingent liability with fiscal implications.
• Governments may not have funds to cover hedging costs, which can range from 7 to 12 percent of the price level protected.
• Governments are often reluctant to make hedging decisions, because they are vulnerable to ex post criticism (and associated political risk).
• There is a lack of technical capacity to manage hedging programs in many countries.

Source: Dana, Rohrbach, and Syroka 2007.

BOX 1.10   Linking changes in productivity and climate to poverty: the use of Envisage and GIDD for long-term scenario building

The long-term scenarios described in this chapter are based on the World Bank's Envisage model with a dynamic core that is essentially a neoclassical growth model. Aggregate growth is driven by assumptions regarding the growth of the labor force, savings and investment decisions (and therefore capital accumulation), and productivity.

The Envisage model has a considerably developed structure (see van der Mensbrugghe 2010 for a detailed description of the model). First, it is multisectoral, which allows for complex productivity dynamics including differentiating productivity growth between agriculture, manufacturing, and services and picking up the changing structure of demand (and therefore output) as growth in incomes leads to a relative shift into manufactures and services. Second, it is linked multiregionally, allowing for the influence of openness—through trade and finance—on domestic variables such as output and wages. The model is also global, with global clearing markets for goods and services and balanced financial flows. Third, the Envisage model has a diverse set of productive fac-

tors, including land and natural resources (in the fossil fuel sectors), and a split between unskilled and skilled workers.

Finally, the Envisage model has been developed into an integrated assessment model with a fully closed loop between economics and climate change. Economic activity generates greenhouse gas emissions. The Envisage model accounts for the so-called Kyoto gases—carbon (C or CO_2), methane (CH_4), nitrous oxide (N_2O), and the fluoridated gases (F-gases). Greenhouse gas emissions are added to the existing stock of atmospheric gases, which also interact with terrestrial and oceanic stocks, leading to changes in atmospheric concentration. The changes in atmospheric concentration convert into changes in radiative forcing that in turn drive changes in atmospheric temperature. The Envisage model closes the loop between the climate and the economy by converting the climate signal as summarized by the global mean temperature into an economic impact.

The Envisage model has a 2004 base year and relies on the Global Trade Analysis Project (GTAP)

price spikes. If the declining responsiveness of demand with per capita income growth is not offset by higher supply responsiveness, than the amplitude of a price spike during a shock will likely be higher. Policy responses matter; they can either amplify or dampen price spikes and either prevent or increase the likelihood of price spikes.

Increases in yields and improved climate resilience, particularly in low-income countries, would reduce the average increase in food prices, the likelihood of price spikes, and the poverty impact of shocks that do happen. Improved agricultural productivity is critically dependent on government support for infrastructure, research that leads to improved climate resilience, and extension, as well as on the establishment of an incentive framework that encourages private

investment and facilitates access to finance for agriculture. To illustrate the potential impact of these improved policies, we develop two scenarios: a baseline scenario consistent with official forecasts; and an alternative scenario that involves a doubling of agricultural productivity growth in developing countries relative to the base line, to about 2 percent annually (as estimated by Martin and Mitra 1999). Both scenarios take into account the consequences of growth and productivity enhancements on climate change and vice versa. The rise in productivity in the alternative scenario reduces international cereal prices by an average of 4 percentage points below base-case levels. As compared to the base line, global agricultural output would increase by another 7 percentage points and global cereal production by also an additional

BOX 1.10   Linking changes in productivity and climate to poverty: the use of Envisage and GIDD for long-term scenario building (continued)

database to calibrate initial parameters. Productivity is derived by a combination of factors. First, agricultural productivity is aligned with the International Food Policy Research Institute's model assumptions of agricultural productivity, that are based on country- and crop-specific crop modeling using the IMPACT model. At the world level, the average growth in productivity over the next 15 years is projected to be around 1 percent a year, about half the long-run recent historical average (see Martin and Mitra 1999). The regional variation is somewhat narrower than in the past, with the highest productivity growth in the Middle East and North Africa followed by Sub-Saharan Africa. Productivity growth in manufacturing and services is labor-augmenting only (both unskilled and skilled). The two are linked with productivity in manufacturing, which is assumed to be higher than in services. The Envisage model assumes that energy efficiency improves autonomously by 1 percent a year in all regions and that international trade costs decline by 1 percent a year.

The Global Income Distribution Dynamics (GIDD), a global computable general equilibrium

microsimulation model, takes into account the macroeconomic nature of growth and of economic policies and adds a microeconomic—that is, a household and individual—dimension. The GIDD includes distributional data for 121 countries and covers 90 percent of the world population. It is used to assess growth and distribution effects of global policies such as multilateral trade liberalization, changes in agricultural productivity, and policies dealing with climate change, among others. The GIDD also allows an analysis of the impact on global income distribution of different global growth scenarios and distinguishes changes resulting from shifts in average income between countries from changes attributable to widening disparities within countries.

The macro-micro modeling framework described here, that is, the combination of Envisage and GIDD, takes into account the consequences of the policy simulations with Envisage on the global income distribution with GIDD, so as to estimate their impact on global poverty.

Source: van der Mensbrugghe 2010.

2 percentage points, relative to their respective 2025 outcomes in the base line (box 1.10 provides a description of the model used).

Faster productivity growth in developing countries helps many net food importers.[12] For example, in the alternative scenario, Sub-Saharan Africa would become self-sufficient in cereals production by 2025, as would Latin America and the Caribbean and Europe and Central Asia (figure 1.5). The Middle East and North Africa region would decrease its dependence on imports of cereals. Only East Asia and the Pacific and high-income countries would experience a drop in self-sufficiency in cereals production.

Higher productivity in agriculture in conjunction with climate change reduces overall poverty further but not in all regions (table 1.9). Given the larger percentage of

FIGURE 1.5   Ratio of cereal production to consumption in 2010 and 2025

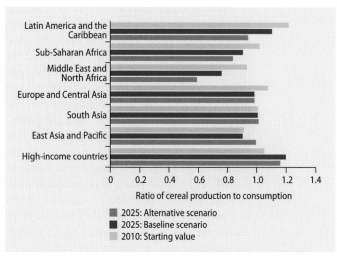

Ratio of cereal production to consumption

- 2025: Alternative scenario
- 2025: Baseline scenario
- 2010: Starting value

Source: World Bank Envisage model.

TABLE 1.9 Poverty forecast, 2015–25
Percent of population living on less than $1.25 a day, 2005 PPP

Region	2015	2025 Baseline, including climate change	2025 Doubling of productivity in developing countries
East Asia and Pacific	7.7	3.0	3.1
Eastern Europe and Central Asia	0.3	0.2	0.1
Latin America and the Caribbean	5.5	5.3	5.4
Middle East and North Africa	2.7	2.3	2.1
South Asia	23.9	14.8	11.8
Sub-Saharan Africa	41.2	34.8	33.2
Total	16.3	12.1	10.8

Source: Up to 2015: World Bank staff calculations from PovcalNet database; for 2025: Envisage and GIDD.

population active in agriculture in Africa and South Asia, the poverty headcount is reduced in these regions by 1.6 and 3.0 percentage points, respectively, taking possible adverse effects of increased agricultural productivity on climate change into account. Poverty increases marginally in East Asia and Pacific and in Latin America and the Caribbean because fewer people are dependent on agriculture, so increases in productivity do little to reduce poverty, and the adverse implications of climate change affect these regions more than elsewhere. Latin America and the Caribbean is expected to be affected by a reduction in tourism revenues, while East Asia and the Pacific could face additional water stress (van der Mensbrugghe 2010).

These scenarios are intended to illustrate the central role of increasing productivity in limiting food price increases. The projected productivity growth may not be achieved for numerous reasons, such as more-stringent-than-expected limits on the availability of productive land, the uncertainty concerning the impact of climate change, and the potential lack of public investment and incentive framework that encourages private investments. Nevertheless, the scenarios do serve to underline the importance of government policies that support increased productivity, both in establishing an appropriate framework to encourage private investment and in providing direct support to the agricultural sector.

Notes

1. The World Bank Agriculture Price Index includes the food price index, plus cocoa, coffee, tea, cotton, jute, rubber, tobacco, and wood.

2. Focus groups and interviews were carried out in 17 countries with respondents representing groups exposed to economic shocks, such as workers in export-oriented sectors, informal sector workers, and farmers. The research explored to what extent and by what means people were able to remain resilient against the recent economic shocks. The data is based on 13 countries for which the qualitative data permitted the authors to determine the importance of these coping responses. The countries were Bangladesh, Cambodia, Central African Republic, Ghana, Kazakhstan, Kenya, Mongolia, Philippines, Serbia, Thailand, Ukraine, Vietnam, and Zambia. See Heltberg, Hossain, and Reva forthcoming.

3. Water use projections to 2050 suggest that the water supply to some 47 percent of the world's population, mostly in developing countries, will be under severe stress, largely because of developments outside of agriculture (OECD-FAO 2011).

4. Both the FAO and the U.S. Department of Agriculture publish stock-to-use estimates. They reflect the difference between estimated production and carry-over stocks on the one hand, and estimated consumption and trade on the other. The stock-to-use measure thus includes (conceptually) all commercial, public, and household stocks, whether or not the stocks in question are actually available for international sale.

5. Although Kazakhstan is located in Central Asia, for grain exports it is often said to belong to the Black Sea region because it uses the seaport facilities in Russia and Ukraine for overseas exports.

6. While export bans imposed by larger exporting countries with a readily available surplus have a greater impact on global prices than export bans imposed by small producers, all export bans can affect markets by leading to a perception of larger-than-actual shortages

and could result in beggar-thy-neighbor actions.

7. Synergies should be explored with the monitoring of the social and poverty impacts of crisis in real time that serves social assistance provision and other support.

8. The AMIS and the associated Rapid Response Forum were launched by the French Presidency of the G-20 in Rome on September 15–16, 2011. The Secretariat is housed at the FAO in Rome. The participants of AMIS are the G-20 countries, Spain, and seven developing countries, which together account for more than 90 percent of world food production and consumption.

9. Biofuel production through crops, like sugar cane, that do not directly compete with food consumption, is likely to have hardly any (or no) impact on food prices.

10. The United States abolished tax credits and import duties for ethanol in December 2011.

11. The promotion of the use of biofuels by some governments has been driven in part by the intention to reduce dependence on fuel imports and generate environmental benefits by replacing oil-based fuel with biofuels.

12. The impact on agricultural productivity of climate change has been widely studied and debated, and the results presented are surrounded by a significant amount of uncertainty.

References

Ahmed, S., and H. G. P. Jansen, eds. 2010. *Managing Food Price Inflation in South Asia.* Dhaka: University Press Limited.

Alderman, H., and T. Haque. 2006. "Countercyclical Safety Nets for the Poor and Vulnerable," *Food Policy* 31, no. 4: 372–83.

Baffes, J. 2010. "More on the Energy/Nonenergy Price Link." *Applied Economics Letters* 17: 1555–58.

Baker, J. 2008. *Impacts of Financial, Food, and Fuel Crises on the Urban Poor.* Washington, DC: World Bank.

Breisinger, C., O. Ecker, and P. Al-Riffai. 2011. "Economics of the Arab Awakening: From Revolution to Transformation and Food Security." IFPRI Policy Brief 18, International Food Policy Research Institute, Washington, DC.

Christiaensen, L. 2007. "Special Focus: Agriculture for Development, East Asia and Pacific Update." World Bank, Washington, DC (November).

———. 2011. "Rising to the Rice Challenge: A Perspective from East Asia." World Bank, Poverty Reduction and Economic Management, East Asia and Pacific, Washington, DC.

Dana, J., D. Rohrbach, and J. Syroka. 2007. "Risk Management Tools for Malawi Food Security." Internal policy note, World Bank, Washington, DC.

Dawe, D. 2008. "How Recent Increases in International Cereal Prices Been Transmitted to Domestic Economies? The Experience in Seven Large Asian Countries." UN FAO-ESA Working Paper 08-03, Rome.

de Hoyos, R., and Medvedev, D. 2011. "Poverty Effects of Higher Food Prices: A Global Perspective." *Review of Development Economics* 15, no. 3: 387–402.

Deininger, K., D. Byerlee, J. Lindsay, A. Norton, H. Selod, and M. Stickler. 2011. "Rising Global Interest in Farmland: Can It Yield Sustainable and Equitable Benefits?" World Bank, Agriculture and Rural Development, Washington, DC.

Demeke, M., G. Pangrazio, and M. Maetz. 2009. "Country Responses to the Food Security Crisis: Nature and Preliminary Implications of the Policies Pursued." FAO Initiative on Soaring Food Prices, Rome.

Dorosh, P. 2009. "Price Stabilization, International Trade and National Cereal Stocks: World Price Shocks and Policy Response in South Asia." *Food Security* 1: 137–49.

FAO (Food and Agriculture Organization). 2011a. "Food Prospects and Crop Situation." Rome (October).

———. 2011b. "Guide for Policy and Programmatic Actions at Country Level to Address High Food Prices." FAO's Initiative on Soaring Food Prices, Rome.

———. 2011c. "Save and Grow: A Policymakers' Guide to the Sustainable Intensification of the Smallholder Crop Production." Rome.

G-20 (Group of 20). 2011. "Price Volatility in Food and Agricultural Markets: Policy Responses." Policy Report for G-20 coordinated by the FAO and the Organisation for Economic Co-operation and Development, May 3.

Heltberg, R., N. Hossain, and A. Reva, eds. 2012, forthcoming. *Living through Crises: How the Food, Fuel, and Financial Shocks Affect the Poor.* Washington, DC: World Bank.

Hossain, N., and D. Green. 2011. "Living on a Spike: How Is the 2011 Food Crisis Affecting Poor People?" Oxford: Oxfam.

Ivanic, M., and W. Martin. 2008. "Implications of Higher Global Food Prices for Poverty in Low-Income Countries." *Agricultural Economics* 39: 405–16.

———. 2012a, forthcoming. "Estimating the Short-Run Poverty Impacts of the 2010 Surge in Food Prices." *World Development.*

———. 2012b, forthcoming. "Short- and Long-Run Impacts of Food Price Changes on Poverty." Policy Research Working Paper, World Bank, Washington, DC.

Jones, D., and A. Kwiecinski. 2010. "Policy Responses in Emerging Economies to International Agricultural Commodity Price Surges." Food, Agriculture and Fisheries Working Paper 34, Organisation for Economic Co-operation and Development, Paris.

Kostandini G., B. Mills, and E. Mykerezi. 2011. "Ex Ante Evaluation of Drought-Tolerant Varieties in Eastern and Central Africa." *Journal of Agricultural Economics* 62, no. 1: 172–206.

Martin, W., and D. Mitra. 1999. "Productivity Growth and Convergence in Agriculture and Manufacturing." Policy Research Working Paper 2171, World Bank, Washington, DC.

Minot, N. 2010. "Transmission of World Food Price Changes to African Markets and its Effects on Household Welfare." Paper prepared for the COMESA Policy Seminar "Food Price Variability: Causes, Consequences and Policy Options," Maputo, January 25–26.

———. 2011. "Food Price Volatility and the Management of Public Grain Stocks in Eastern and Southern Africa." Report prepared for the World Bank, Washington, DC (November).

OECD-FAO. 2011. *Agricultural Outlook 2011–2020.* Paris and Rome.

PBL (Netherlands Environmental Assessment Agency). 2010. "Rethinking Global Biodiversity Strategies: Exploring Structural Changes in Production and Consumption to Reduce Biodiversity Loss." Bilthoven, the Netherlands.

———. 2012, forthcoming. "Making Ends Meet: Pathways to Reconcile Global Food, Energy, Climate and Biodiversity Goals. Bilthoven." the Netherlands.

Robles, M. 2011. "Price Transmission from International Agricultural Commodity Markets to Domestic Food Prices: Case Studies in Asia and Latin America." International Food Policy Research Institute, Washington, DC.

Steenblik, R. 2007. "Biofuels: At What Cost? Government Support for Ethanol and Biodiesel in Selected OECD Countries." The Global Subsidies Initiative of the International Institute for Sustainable Development, Geneva.

Stigler, M., and A. Prakash. 2011. "The Role of Low Stocks in Generating Volatility and Panic." In *Safeguarding Food Security in Volatile Global Markets,* ed. A. Prakash, pp. 319–33. Rome: FAO.

Subervie, J. 2008. "The Variable Response of Agricultural Supply to World Price Instability in Developing Countries." *Journal of Agricultural Economics* 59, no. 1: 72–92, 102.

Trostle, R., D. Marti, S. Rosen, and P. Westcott. 2011. "Why Have Food Commodity Prices Risen Again?" USDA Economic Research Service WRS-1103, Washington, DC.

van der Mensbrugghe, D. 2010. "The Environmental Impact and Sustainability Applied General Equilibrium (ENVISAGE) Model, Version 7.1." Technical Reference Document, World Bank, Washington, DC (December).

WFP (World Food Programme). 2008. "Vouchers and Cash Transfers as Food Assistance Instruments: Opportunities and Challenges." Rome.

World Bank. 2007. *World Development Report 2008: Agriculture for Development.* Washingtom, DC: World Bank.

———. 2009. "Implementing Agriculture for Development." WBG Agriculture Action Plan 2010–2012. Washington, DC.

———. 2011a. *Building Resilient Safety Nets: Proceedings of the Social Protection South-South*

Learning Forum. Washington, DC: World Bank.

———. 2011b. *Global Economic Prospects: Maintaining Progress and Turmoil.* vol. 3. Washington, DC: World Bank.

———. 2011c. "MNA Facing Challenges and Opportunities: Regional Economic Update." World Bank, Washington, DC.

———. 2011d. "Responding to Global Food Price Volatility and Its Impact on Food Security." Development Committee Paper, World Bank, Washington, DC (April).

———. 2011e. *World Development Report 2012: Gender Equality and Development.* Washington, DC: World Bank.

———. 2012a, forthcoming. "High Food Prices: Latin America and the Caribbean Responses to a New Normal." World Bank, Washington, DC.

———. 2012b, forthcoming. "Responding to Higher and More Volatile World Food Prices."

Economic Sector Work, Agriculture and Rural Development Department, Washington, DC.

———. 2012c, forthcoming. "Using Public Stocks for Food Security." Economic Sector Work, Agriculture and Rural Development Department, Washington, DC.

———. 2012d, forthcoming. "Greening Growth: A Path toward Sustainable Development." Washington, DC.

World Bank and International Monetary Fund. 2010. "How Resilient Have Developing Countries Been during the Global Crisis?" Development Committee Paper DC 2010–0015, Washington, DC.

Wright, B. 2009. "International Grain Reserves and Other Instruments to Address Volatility in Grain Markets." Policy Research Working Paper 5028, World Bank, Washington, DC.

Zurayk, R. 2011. *Food, Farming, and Freedom: Sowing the Arab Spring.* Charlottesville, VA: Just World Books.

2

Nutrition, the MDGs, and Food Price Developments

Summary and main messages

Even temporarily high food prices affect the long-term development of children. Conditions of early life (from conception to two years) provide the foundations for adult human capital. Vicious interactions between malnutrition,[1] poor health, and impaired cognitive development set children on lower development paths and lead to irreversible changes.

Seemingly small shocks can exert great damage if they are not dealt with early. The most dramatic effect of the food price crisis is an increase in infant mortality, especially in low-income countries. Other hard-to-reverse impacts include growth faltering (stunting or low height for age) and lower learning abilities. Malnourished young children are also at more risk for chronic diseases such as diabetes, obesity, hypertension, and cardiovascular disease in adulthood. Moreover, declines in human capital in a crisis tend to be more pronounced than the corresponding increases during economic booms.

The most vulnerable bear the brunt of the adverse impacts of high food prices, through malnutrition. Poor households tend to spend a larger share of their income on food and are especially vulnerable to price increases.

The dynamics of intrahousehold distribution combined with biological vulnerability mean that pregnant women and children in these households face higher risks of malnutrition. Impacts such as mortality and school dropout are often sharper for girls than boys.

To build household and individual resilience and mitigate long-term effects, interventions can work through multiple pathways, beyond trying to keep prices low. In the short term, interventions should focus on maintaining household purchasing power and caloric and micronutrient intakes through cash transfers, food and nutrient transfers, and workfare-with-nutrition. To maximize impacts on children and women, interventions should ensure that those transfers are put in the hands of women, if possible. In the longer term, interventions should focus on strengthening the link between smallholder agriculture and nutrition, addressing seasonal deprivation, and promoting women's income and girls' education.

Specific interventions need to target vulnerable children through behavioral changes related to breastfeeding, feeding during illness, hygiene, access to micronutrients, deworming (which increases absorption of micronutrients), and preventive and therapeutic feeding. Consequently, activities for countries to mitigate the potentially negative impacts of food prices include improving data quality about nutrition status (height for age, weight for age, and micronutrient deficiencies), practices (breastfeeding), and interventions; targeting the period from conception to two years of life (pregnant women and young children); expanding Scaling Up Nutrition interventions; tailoring interventions to country capacity—in the government, civil society, and private sector—and to country nutrition security issues; and incorporating nutrition-sensitive approaches in multisectoral programs (social protection, health, agriculture, and income-generation interventions).

How high food prices affect the MDGs

Higher food prices may make it more difficult to achieve most Millennium Development Goals (MDGs). Food price increases affect food consumption, quality of one's diet, access to social services, and sometimes the quality of care for infants and young children. All these factors may increase undernutrition among children (and decrease their learning capacity and survival rates), adult women (and if they are pregnant, increase their chances of maternal mortality and affect fetal growth and future outcomes), and adult men (and affect their productive capacity). In addition, undernutrition decreases the efficacy of treatments for HIV/AIDS and other major diseases. Box 2.1 summarizes the combined impact of the food price crisis and malnutrition on the MDGs. Conservative estimates from Grantham-McGregor and others (2007) suggest that over 200 million children under five years of age living in developing countries fail to reach their cognitive development potential because of risks linked to poverty, poor health and undernutrition and lack of stimulation at home. Save the Children (2011) estimates that the recent food price hike put 400,000 children's lives at risk.

How food prices affect nutrition

Food security and nutrition security are different but interlinked concepts. Food security, an important input for improved nutrition outcomes, is concerned with physical and economic access to food of sufficient quality and quantity in a socially and culturally acceptable manner.[2] Nutrition security is an outcome of good health, a healthy

---

BOX 2.1   Impact of higher food prices and undernutrition on the MDGs

- As food prices increase, the purchasing power of the poor decreases, the composition of their diet worsens, and their food consumption may decrease. These changes directly affect all targets of MDG 1 on poverty, full and productive employment, and hunger.
- Malnutrition affects early childhood development and makes children more likely to drop out of school (MDG 2).
- An increase in food prices affects women and girls' consumption disproportionately (MDG 3).
- Undernutrition is linked directly to more than one-third of children's deaths each year (MDG 4).

- Pregnant women face heightened maternal mortality, through increased anemia, during a food price crisis (MDG 5).
- The adverse effects of a food crisis on the availability of health services and on health status bear on countries' and individuals' abilities to combat the HIV/AIDS epidemic (MDG 6).
- Undernutrition weakens the immune system and compounds the effect of diarrhea and waterborne diseases (MDG 7).
- Higher food prices have weakened intergovernmental coordination in food markets (MDG 8).

environment, and good caring practices as well as household food security (World Bank 2006). For example, a mother may have reliable access to the components of a healthy diet, but because of poor health or improper care, lack of knowledge, gender, or personal preferences, she may be unable, or choose not, to use the food in a nutritionally sound manner, thereby becoming nutritionally insecure. Nutrition security is achieved for a household when secure access to food is coupled with a sanitary environment, adequate health services, and knowledgeable care to ensure a healthy life for all household members. A household (or country) may be food secure, yet have (many) individuals who are nutritionally insecure.

Food security is therefore a necessary but not sufficient condition for nutrition security. And although households make key decisions that influence the nutritional status of their individual members, government funding and policy decisions determine the environment in which households operate (IFAD, WFP, and FAO 2011).

Nutrition security is multidimensional. Solutions to improve nutrition in a given country environment will require integration among the sectors most relevant to individuals' nutritional status, such as trade and infrastructure, agriculture, and the labor market, as well as the social sectors such as health, education, and social protection (Ecker, Breisinger, and Pauw 2011). A shock such as the food price crisis affects both household and government behavior.

Effects at the household and individual levels

Dietary quality and food quantity may be affected

As prices rise, households will first try to replace pricier foods with cheaper sources of calories, moving from some food categories or shifting to lower-quality foods. When prices increase further and substitution is not enough, households decrease their caloric consumption. In the first adjustment, poor consumers shift from foods such as meats, fish, vegetables, and fruits to staple foods, such as cereals and tubers, and their protein and micronutrient intake may suffer. Young children—in utero and during their early years—who have high nutrient needs for iron, vitamin A, and zinc, among others, may be particularly at risk and will bear long-term impacts of this "hidden hunger." In the second adjustment, households decrease their caloric consumption—urban households in Pakistan (Friedman, Hong, and Hou 2011) and poor households in Haiti (World Bank 2010b), for example—and the number of children with low weight for age increases. In Vietnam Gibson and Kim (2011) show that a 10 percent increase in the relative price of rice reduces calories by less than 2 percent, but they estimate that this elasticity would be more than 4 percent if they ignore substitution into lower-quality rice, as households in Vietnam protect calorie consumption by downgrading the quality of their food intake.

In urban areas, street foods are central to food consumption patterns among the urban poor. In Accra and in Latin America, street food may account for nearly 40 percent of the total food budget of the urban poor (Ruel 2000). The risk from higher food prices is an increase in consumption of street foods, which are rich in oil and starch. This results in diets of high energy density (caloric content) and little nutritional value, contributing to already rising obesity rates among the urban poor, as in Mexico (CONEVAL 2009) the United States (Centers for Disease Control and Prevention 2011), and in many middle-income countries undergoing the nutrition transition from high levels of undernutrition to overnutrition.

Women and children may have to increase their workforce participation

Increased women's labor force participation may yield positive results on household income and purchasing power, but it is likely to change childcare arrangements. The effect of mother's increased workforce participation on child welfare depends on children's ages, other household resources, and the education and knowledge of the person responsible for childcare and feeding. In noncrisis settings, in urban poor communities of Guatemala City and Accra, mothers seemed to be able to manage their childcare responsibilities and their income-generating role efficiently (Levin et al. 1999, Ruel et al. 1999, 2002). But in crisis settings, if women engage in distress work (work in response to an adverse shock to the main earner's income) such as they do in rural India (Bhalotra 2010), time constraints decrease time spent seeking health care, and infant girl mortality may increase. (Rural households where mothers are uneducated or had a first birth as a teenager are driving these results.) Interventions that address women's childcare and pregnancy needs (such as crèches around temporary construction sites in India) can help to protect children's well-being.

The effect of high food prices on children's labor force participation is ambiguous. Children may join in productive agricultural activities if the household feels it cannot afford schooling any more. Children who drop out of school find it difficult to return to school when the crisis is over, and their schooling attainment suffers. Children's income may also become a key contribution to maintaining the household's caloric intake. If, though, the price crisis is also a jobs crisis, as in Europe and Central Asia in 2008 or in Peru in 1988–92 (Schady 2002), children may not increase their workforce participation.

If households seek less health care or the supply of health services decreases, individual members' health may deteriorate and affect their nutritional status

When households feel they cannot afford health care expenses, the health status of adults and children may suffer. Poor health affects nutrition through changes in metabolism, malabsorption of nutrients and appetite loss, and changes in feeding practices. Highly prevalent diseases such as acute respiratory infections and diarrhea reduce the absorption of nutrients such as vitamin A from the small intestine, establishing a vicious cycle because vitamin A deficiency depresses the immune

system and makes the child more susceptible to subsequent infections. Feeding practices, such as decreasing liquid intake of children affected by diarrhea, may also have severe consequences.

The poor bear the brunt of decreases in funding of primary care and community-based nutrition interventions (Alderman 2011b). Latin America's economic crisis of the early 1980s cut public health spending, which had a disproportionate effect on the poorest groups (Musgrove 1987). Ferreira and Schady (2009) contrast the experience of Indonesia and Peru to show the importance of maintaining critical services to avoid increases in child undernutrition during crises. In Peru the crisis caused a collapse in public health expenditures of over 60 percent and declines in health service utilization (including more home births and fewer prenatal checkups). Infant mortality shot up from 50 per 1,000 live births in 1988 to 75 in 1990. In contrast, in Indonesia increased donor aid made up for some of the shortfall in government spending. Infant mortality still spiked from 30 per 1,000 live births in 1996 to 48 in 1998, but nutrition indicators such as wasting, stunting, and anemia did not worsen.

Intrahousehold reallocation and care practices may mitigate or aggravate the effects of food price increases on specific household members

Women often become "shock absorbers of household food insecurity," as they reduce their own consumption to allow for more food for other household members (Quisumbing, Meinzen-Dick, and Bassett 2008). Rural poor women in the United States and Canada (McIntyre et al. 2003) tend to both lower and change their dietary intake in favor of their children (particularly in terms of energy, vitamin A, folate, zinc, calcium, and iron) when they experience food insecurity. In some communities in Bangladesh, Indonesia, Jamaica, Kenya, and Zambia (Holmes, Jones, and Marsden 2009), when choices have to be made, children come first; in other communities, men are favored. In none of the communities, however, were women, including pregnant women, offered the most nutritious foods. In Indonesia mothers buffered children's caloric intake during the 1997–98 crisis, resulting in increased maternal wasting and anemia (Block et al. 2004).

Women's lack of education and low status in the household contribute to child malnutrition, as do poor care practices. Poor child feeding practices are responsible for high levels of undernutrition and affect girls more than boys in most countries in South Asia. In many countries, mothers do not exclusively breastfeed their children during the first six months of life (see below), and the foods used to complement breast milk are often low in energy and essential micronutrients. The knowledge of a grandmother or an older sibling who cares for the child may even be more limited than the mother's. Women's education and status within the household contributed to more than 50 percent of the reduction in child undernutrition between 1970 and 1995 (Quisumbing et al. 2000). Good care practices can mitigate the effects of poverty and low maternal schooling in child nutrition (Armar Klemesu et al. 2000).

Increasing income is not enough

Among households, undernutrition rates can be high even among the food secure. For example, if the lowest two quintiles by wealth in Pakistan had the same characteristics as the third quintile, poverty would be eliminated, but 38 percent of children would still be malnourished. In Ethiopia 40 percent of children in the wealthiest quintile are stunted. This pattern is consistent across many countries (Haddad et al. 2003) and points to the need for interventions beyond general poverty reduction to address specific nutritional issues. As noted by the World Bank (2006), several reasons explain this pattern:

- Pregnant women eat too few calories and too little protein, and have untreated infections, such as sexually transmitted diseases that lead to low birth weight.
- Mothers have too little time to take care of their young children or themselves.

- Mothers of newborns discard colostrum, the first milk, and thus lose the boost to the infant's immune system that colostrum provides.
- Mothers rarely breastfeed infants under six months exclusively, even though breast milk offers the best source of nutrients and protects against many infections.
- Caregivers start introducing complementary solid foods too late. They feed children under age two too little food or foods that are not energy dense.
- Although food is available, intrahousehold food allocation practices may mean that women and young children's energy needs are not met and that their diets are poor in micronutrients or protein.
- Caregivers do not know how to feed children during and following diarrhea or fever.
- Caregivers' poor hygiene contaminates food with bacteria or parasites.

Box 2.2 illustrates some of these effects—on quantity and quality of food consumption, individual workforce participation, intrahousehold allocation, access to services, and other coping mechanisms—in northern Bangladesh during the 2007–08 food price crisis.

Effects at the national level

Increased state spending on food purchases and subsidies can divert resources from health and education (among other sectors), yet these are key sectors for nutritional status, because undernutrition is often linked to preventable diseases (such as diarrhea) and lack of nutrition knowledge (for example, information about optimal feeding practices for infants and young children). As food prices rose, many governments expanded (or set up) food subsidy programs to alleviate economic hardships. In the Middle East and North Africa, for example, spending on these programs reached 5–7 percent of gross domestic product (GDP). But such programs entail trade-offs and may threaten other investments.[3] In addition, price subsidies generally target foods

BOX 2.2   The impact of the 2007–08 food price spike on a rural community in northern Bangladesh

Bangladesh has high levels of child undernutrition (36 percent stunting, 16 percent wasting, and 46 percent underweight). Prices of key staples increased by as much as 50 percent from 2007 to 2008. On top of this, the country suffered floods in mid-2007 and a cyclone in November 2007, which reduced the *aman* (or second) rice harvest. Export restrictions by India, one of the country's main rice providers, also raised rice prices.

An assessment of livelihood and nutrition security in Kurigram village (194 households) in 2005 and a follow-up assessment in November 2008 (250 households) shows some of the effects of the price hike. The richest households benefited from the price hike (as rice producers). One-third to one-half of households had lower disposable income after the crisis, mainly because of the rice price hike and, to some minor extent, crop failure in one of the rice harvests. (Disposable income was taken as cash income left, after households had met their food energy requirements per adult equivalent, using a cost-of-diet approach.)

The poorest quartile was no longer able to afford a diet that provided them with their energy and micronutrient needs. Children ate fewer meals, had less diverse diets, and received few nutrient-rich foods. Stunting among children in the poorest households

was twice as high as in the richest households. A 7 percentage point improvement in stunting rates (probably linked to improvements in women's status and better road infrastructure) was lost during the crisis—a loss that will have permanent consequences for the children's mental and physical development.

Families in the community responded to the price hike by sending children to work, taking children out of school, selling productive assets, and reducing their food intake. Poor families took loans to replace lost income, and repayment became a priority over livelihoods and diet investments. Three families moved to Dhaka, the capital.

Even though the richest households benefited, agricultural labor wages did not rise enough to compensate poorer households for the price rise (partly a result of the *aman* crop failure). Only one household benefited from the government's 100-day rural employment program. No household received subsidized rice, although children in school received food-for-education transfers. Some households benefited from the cereal program, some fertilizer stipends, and some stipends for the elderly, widows, and freedom fighters.

Source: Save the Children 2009.

that are low in micronutrients, distort relative prices, and may create negative incentives for people to diversify their diets once the crisis is over. For example, in Morocco, the subsidy on soft wheat flour is supporting most of the milling sector (World Bank 2005). Expenditures on physical infrastructure and especially roads are not generally considered as important for nutritional status, even though they are key both to establish a food supply chain that moves food from consumers to producers through markets and to enable households' access to health, education, and, to a lesser extent, social assistance services.

The connection between economic growth and poverty reduction is well established, but

the correlation between income growth and nutrition gains is much weaker (Ecker, Breisinger, and Pauw 2011; Headey 2011). Undernutrition countrywide (defined as low weight for age) may decline at very roughly half the rate that per capita gross national product increases (Alderman 2011a)—28 percent in India, 67 percent in China, and 76 percent in Bangladesh in the 1990s. Yet, Deaton (2010) reports that in India per capita calorie consumption fell in 1997–2007, despite high rates of per capita income and consumption growth. While some of the calorie reduction may be linked to less physical activity (as people spend less time in agriculture) or to lower morbidity, the puzzle remains.

The interaction of crises and biology

Short-term shocks, long-term effects

The most pernicious effect of the crisis is an increase in infant mortality, especially girl infants in low-income countries. A recent study by Baird, Friedman, and Schady (2011) shows a large, negative association between declines in per capita GDP and mortality of infants between birth and one year of age. The study, which analyzes data from 59 Demographic and Health Surveys and 1.7 million births, also reveals that the mortality of children born to rural and less educated mothers is more sensitive to economic shocks, which suggests again that the poor bear the brunt of crises. In addition, the mortality of infant girls is significantly more sensitive to income shocks than that of boys. In a companion study, Friedman and Schady (2009) estimate that the 2008 crisis probably led to an excess 35,000–50,000 infant deaths in Sub-Saharan Africa in 2009 and that nearly all these excess deaths were among girls.

Interventions that tackle child mortality benefit country's growth overall. Apart from the moral arguments for tackling child mortality, analysis by Baldacci et al. (2004) and Save the Children (2008) showed that a 5 percent improvement in child survival rates raises economic growth by 0.85 to 1.0 percentage point a year over the following decade.

Less dramatic but also severe are the potentially negative effects of economic crises on nutritional and environmental pathways that influence early childhood development and subsequent life opportunities. These arise from interactions between undernutrition, health, and learning, which set children on lower development paths and lead to changes in states that are difficult to reverse—it is easier, for example, to maintain a child in school than to reenroll once she or he has dropped out. The timing of the crisis in the life cycle also matters, with the period from conception to two years of life being one of high risk because of physical and cognitive development. Nutritional deprivation during that period can cause irreversible setbacks in growth and sociocognitive development (Victora et al. 2008). The accumulation of toxic stress in the first years of life—through decreased care and transitions in and out of poverty—has long-term consequences for an adult's wages and productivity.

Deteriorations of human capital during economic downturns and improvements during booms are asymmetric

The *Global Monitoring Report 2010* reported that human development indicators during downturns tend to worsen more than they improve during economic booms. For example, life expectancy decreases by 6.5 years during decelerations but may increase by only 2.0 years during growth accelerations. Similarly, the increase in infant mortality during deceleration is three times the decrease during accelerations (24 versus 8 per 1,000 live births), and the decrease in primary schooling completion rates during deceleration is six times the increase during acceleration (25 percent versus 4 percent). Undernutrition contributes to more than a third of infant deaths and decreases learning abilities and school attainment (see below).

Economic downturn affects girls more than boys. Life expectancy decreases by seven years for girls and six years for boys during bad times (it increases by two years for both during good times). Primary education completion rates fall by 29 percent for girls and 22 percent for boys during bad times and rise by 5 percent for girls and 3 percent for boys during good times. Female-to-male enrollment ratios fall severely during downturns, with higher drops in tertiary and secondary education than in primary education.

Large scale and extreme shocks cause increases in low birth weight, wasting, and stunting

During Argentina's crisis in 1999–2002, the elasticity of low birth weight to GDP was −0.25 cases per 1,000 births (Cruces, Gluzmann and Lopez Calva 2010). Stunting increases as a result of extreme shocks, such as the drought in Zimbabwe in 1994–95

(Hoddinott and Kinsey 2001), crop damage in Ethiopia in 1995–96 (Yamano, Alderman, and Christiansen 2003), and the very large economic contractions experienced in Peru in 1988–92.

Impacts from more moderate crises are heterogeneous

Such impacts include increased underweight and anemia and decreased access to health services. In Cameroon the share of underweight children under age three increased from 16 percent in 1991 to 23 percent in 1998 as a result of combined economic crises and subsequent government adjustment programs (Pongou, Salomon, and Ezzati 2006). Declines in economic status and health care accessibility were both correlated with an increase in undernutrition in urban areas. In rural areas, reductions in health access were correlated with an increase in undernutrition, especially among children born to little-educated mothers or poor households. It is unclear, however, whether the lower access stemmed from weakened ability to pay or from reduced provision of health services. In Central Java in 1997–98, drought and financial crisis were associated with a decrease in mean iron hemoglobin concentration of 6.1 percent and increasing anemia, with larger effects on children born or conceived during the crisis (Waters, Saadah, and Pradhan 2003). The latter suggests that maternal undernutrition was an additional risk pathway, which is consistent with decreases in consumption of green leafy vegetables, eggs, and cooking oil among households.

The importance of when: window of risk and opportunity from conception to 24 months[4]

Early life conditions have a disproportionate influence on forming adult human capital, understood in terms of height, skills (cognitive and noncognitive) and capabilities (such as health and social functioning) (Victora et al. 2008; Friedman and Sturdy 2011). A particularly critical period for brain development is from the first few weeks in the womb to the second year of life. Early cognitive and sensory-motor development, as

well as socioemotional competence, affect school preparedness and subsequent school performance. This is also a period of intense physical development: children are expected to grow 50 centimeters in utero, 24 centimeters in their first year of life, and 12 in their second year, after which time the pace of growth slows until the rapid growth spurt of adolescence. Figure 2.1 shows how the shortfall between children in different regions remained unchanged after 24 months, when compared with a healthy reference group.

Risk factors that affect children in low-income countries include intrauterine growth restriction (11 percent of births), stunting (around one-third of children under five years), iron deficiency (one-fourth to one-third of children under four years), iodine deficiency (one-third of the population worldwide), maternal depression (one-sixth of postpartum mothers), and inadequate cognitive stimulation (Friedman and Sturdy 2011). Iron deficiency is associated with fetal and child growth failure, lower cognitive development in children, lower physical activity and productivity in adults, and increased maternal mortality. Vitamin A deficiency causes blindness and is a risk factor for increased severity of infections, which leads to increased mortality. Zinc deficiency is associated with stunting and higher incidences of diarrhea and pneumonia. Iodine deficiency affects cognitive development and reduces intelligence (IQ). Lower quantity and quality of nutritional intake, lower household income, lower state resources, and lower quality care would cause increases in the prevalence of low birth weight, childhood wasting, and then stunting, which in turn have significant negative impacts on children's development.

The double burden of malnutrition and chronic disease[5]: Malnourished children may become overweight adults

Many aspects of fetal growth influence long-term health, and children who experienced malnutrition in utero and in their early years are more at risk of chronic diseases such as type 2 diabetes, abdominal obesity, hypertension, and cardiovascular disease. For example, children in utero during the famine in the Netherlands of 1944–45 show increased risk of chronic disease and mental illness in middle age and greater loss of attention and cognitive ability than the general population as they age further (Alderman 2011a). Similarly in India, children who were thinner in infancy and experienced rapid growth show a higher prevalence of diabetes (box 2.3), giving that country the highest numbers in the world, both of malnourished children and of people with diabetes. Many countries in Latin America face increases in overweight and obesity among adults who were previously undernourished, as well as high numbers of chronically undernourished children (figure 2.2). During the nutritional transition from under- to overnutrition, children who were undernourished face higher risks of overweight and obesity as adults, while at the same time, lack of nutritional knowledge and poverty with micronutrient poor diets still undermine the development of children.

FIGURE 2.1   Mean height for age (Z-scores) by age, relative to WHO standards, by region

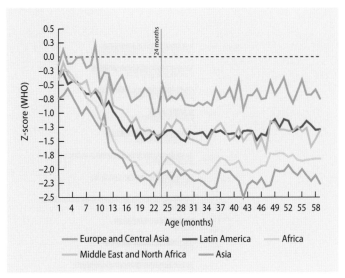

Source: Victora et al. 2010.
Note: Europe and Central Asia countries included are Armenia, Kazakhstan, Kyrgyzstan, Moldova, Mongolia, Montenegro, and Turkey. Latin America countries included are Bolivia, Brazil, Colombia, Dominican Republic, Guatemala, Haiti, Honduras, Nicaragua, and Peru. Middle East and North Africa countries included are Arab Republic of Egypt, Jordan, Morocco, and Republic of Yemen. Thirty Sub-Saharan African countries are included. Asia countries included are Bangladesh, Cambodia, India, and Nepal.

BOX 2.3   Malnutrition and chronic disease in India

Some 42 percent of the 160 million children in India under the age of five are underweight. Prime Minister Manmohan Singh described the situation as a matter of "national shame" and undernutrition as "unacceptably high" when he announced those numbers in January 2012. There are signs of progress—one in every five children has reached an acceptable healthy weight over the past seven years in 100 focus districts, which were particularly badly off. But the current figures point to the inadequacies and inefficiencies of government initiatives (such as the Integrated Child Development Scheme), the scale of the needs of India's child population (the largest in the world), and the lack of awareness about nutrition. And those numbers may be only the starting point of a much larger long-term problem.

A longitudinal study of a cohort of births in South Delhi followed to age 32 found that those children who were thinner in infancy—with a body mass index (BMI) under 15—had an accelerated increase of BMI until adulthood. Although none was classified as obese by age 12, those with the greatest increase in BMI by this age had impaired glucose tolerance or diabetes by the age of 32 (Bhargava et al. 2005, cited in Alderman 2011a). Similar results have been reported using a panel in Pune (Yajnik 2009, cited in Alderman 2011a).

The transition from a resource-poor environment to one that is less constrained may aggravate these risks. India has not only the largest number of undernourished children in the world, it also has the most people with diabetes (Ramachandran and Snehalatha 2010). These two statistics may very well have a common origin. While the Indian population does not have a high rate of obesity relative to the rest of the world, there is a tendency to accumulate adipose tissue around the waist. This pattern is associated with elevated risk of chronic disease.

FIGURE 2.2 Percentage of stunted children and overweight women in selected Latin American countries

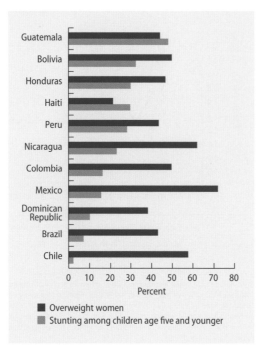

■ Overweight women
■ Stunting among children age five and younger

Source: World Health Organization global databases on child growth and malnutrition and on body mass index.
Note: Overweight: body mass index greater than 25; stunting: height for age less than two standard deviations using National Center for Health Statistics reference.

In times of high food prices, the double burden increases and obesity and undernutrition may coexist within the same household and the same person. As mentioned, poor families switch away from nutritious food and buy "empty calories," as is happening in Honduras and Guatemala (Robles and Torero 2010). Combined with the changes in metabolism described in the previous paragraph, these empty calories will increase the rates both of stunting and anemia and of overweight and obesity in many middle-income countries. In the Arab Republic of Egypt, Peru, and Mexico, about half the women with anemia are overweight or obese.

One possible explanation for the impact of fetal growth on disease later in life is linked to adapting to stress in the womb. The signal derived from limited nutrients in utero may lead to an adaptation in which the child becomes particularly efficient at conserving resources. However, if that individual is subsequently confronted with a resource-rich environment, this maladapted response contributes to overnutrition and increased risk of chronic disease. It may also threaten the welfare of the next generation because hyperglycemia or diabetes in mothers increases the risk of diabetes for their offspring (Delisle 2008).

Crises may be transitory events, but their impacts on young children are not—they continue in the medium term unless stemmed by interventions. Poor children who were under age three during Ecuador's 1998–2000 crisis showed increased stunting and lower vocabulary test scores (a measure of cognitive development) in 2005 when they were five to seven years old (Hidrobo 2011). This finding suggests that they may have experienced reduced parental time on care, and their households may not have managed to protect them from the general health environment deterioration caused by El Niño and cuts in public services. Rural farming households and households with access to health centers were better able to protect the height of their children but not their vocabulary score.

Where interventions were in place, nutritional status improved. In Senegal, the national nutrition program adopted community-based approaches, targeted the "first 1,000 days," implemented systematic nutrition screening, and delivered interventions using a network of well-supervised nongovernmental organizations (NGOs) (Alderman et al. 2008). Over the years, the program added bednet distribution, community management of acute undernutrition and food fortification, and, most recently, a cash transfer initiative. Prenatal care increased from one-third to two-thirds, exclusive breastfeeding for the first six months doubled to 58 percent, and correct use of bednets more than doubled to 59 percent. The rate of stunting in 2005 represented just 59 percent of that in 1990. Similarly, the underweight rate in 2005 was 65 percent of that in 1990.

Childhood exposure to adversity (both extreme events such as drought, civil war,

---

BOX 2.4   Consequences of early childhood growth failure over lifetimes in Guatemala

Growth failure in early life in rural Guatemala, as measured by low height for age (stunting) at 36 months, affects a wide range of adult outcomes: education, choice of marriage partners, fertility, health, wages and income, and poverty and consumption. The data are based on interviews between 2002 and 2004 of participants in a nutrition supplementation trial between 1969 and 1977.

Participants who had received nutritional supplementation (a high-protein energy drink with multiple micronutrients) and free preventive and curative medical care (including the services of community health workers and trained midwives, as well as immunization and deworming) were less likely to become stunted.

Otherwise, participants who were stunted at 36 months of age left school earlier and had significantly worse results on tests of reading and vocabu-

lary and on nonverbal cognitive ability some 35 years later. They also married people with lower schooling attainment. Women had 1.86 more pregnancies and were more likely to experience stillbirths and miscarriages. No link was found with greater risks of cardiovascular or other chronic disease.

Individuals who were not stunted earned higher wages and were more likely to hold higher-paying skilled jobs or white-collar jobs. They were 34 percentage points less likely to live in a poor household. A one standard deviation increase in height for age lifted men's hourly wage by 20 percent, increased women's likelihood of operating their own business by 10 percentage points, and raised the per capita consumption of households where the participants lived by nearly 20 percent.

Source: Hoddinott et al. 2011.

---

and famine, and more moderate events such as low rainfall) has long-lasting adult life consequences. Exposure to the Chinese Long Walk famine of 1959–62 is associated with higher illiteracy, lack of work, and disability later in life. Findings are similar for the Greek famine of late 1941 and early 1942, with those who were in their first year of life when the famine struck the most affected. Exposure to the famine as a child lowered literacy, upper secondary and technical schooling, and occupational prestige. Even moderate adversity, such as low rainfall during the year of birth, has been associated with reduced child growth and increase child morbidity in India, and decreased adult height and schooling in Indonesia (Maccini and Yang 2009). The clearest pathway through which these impacts occur is nutrition, especially during the critical period of early childhood, from six months of age and beyond when a child transitions to complementary foods in addition to breast milk.

Height for age at two years is the best predictor of human capital, and undernutrition is associated with lower human capital.

Stunted children have poorer performance in school (reduction in test scores equivalent to two years of schooling lost). With the observation that every year of schooling is equivalent to an average increase of 9 percent in adult annual income, Grantham-McGregor et al. (2007) estimate a loss in adult income of between 22 and 30 percent for stunted children. A long-run longitudinal study in Guatemala (Hoddinott et al. 2011) provides additional evidence on long-term impacts of undernutrition, including on body size and adult fitness, and wages and types of employment (box 2.4). While the long-term effects of growth failure are severe, interventions such as nutritional supplementation and basic medical care in the early years have strong potential to improve outcomes over the course of beneficiaries' lives.

Building resilience: interventions to mitigate the effects of the food crisis

Food crises are regular occurrences in developing countries, but interventions to decrease malnutrition can mitigate their impacts. The many pathways along which food crises affect household and individual welfare also offer multiple entry points for interventions. We discuss these entry points and the associated costs under three headings: consumption and social protection, biology and health, and production and income generation. Intervention packages will of course vary by country development and capacity, as well as by the types of problems faced, but there is a broad consensus on the beneficial impact of proven interventions (World Bank 2012 forthcoming).

Consumption and social safety nets

When food prices increase, and before food-output systems can adapt, some safety net interventions seek to maintain consumption in the short run, especially among more vulnerable groups. These interventions may also have longer-term impacts and can contribute to bridging the twin-track approach to food security, promoted by the Food and Agriculture Organization of the United Nations (FAO): short-term transfers and relief to protect consumption and long-term investments to increase food output.

Cash transfers

One response to rising food prices is to support consumption of the poor through targeted cash transfers. Conditional cash transfers have shown some results on nutritional outcomes (Fiszbein and Schady (2009) provide a review). Fernald, Gertler, and Neufeld (2008) report positive impacts of Progresa/Oportunidades on children's height in Mexico,[6] Attanasio et al. (2005) show similar effects of Familias in Acción in Colombia, and Ferrreira et al. (2011) for cash transfers in Brazil. Macours, Schady, and Vakis (2008) provide evidence on the nutritional and early childhood development impacts of a conditional cash transfer pilot designed to address crises such as droughts, cyclones, and extreme poverty in Nicaragua. Payments were conditional on school attendance for school-aged children and on preventive care visits for preschool children. Parents also

received information about nutrition and the importance of food choices. The pilot led to significant gains for a variety of cognitive and noncognitive skills (social and personal measures and vocabulary). The program shifted household expenditure toward more diversified diets, more nutrient-rich foods for young children, and materials offering greater stimulation, such as books and paper. In addition, children benefited from an expanded menu of nutrition interventions in health services, including micronutrient supplementation, growth monitoring and promotion, and deworming. Similarly in Malawi, Miller, Tsoka, and Reichert (2011) report that the unconditional M'chinji social cash transfer enabled beneficiary households to avoid food shortages and increase dietary diversity, and that children were more likely to gain height and report better health (Miller et al. 2010). In Indonesia, Skoufias, Tiwari, and Zaman (2011) show that cash transfers helped protect dietary diversity.

Food and nutrient transfers

Another option to maintain consumption is to transfer food directly to vulnerable households. If inflation is high and erodes the value of cash, some potential beneficiaries may prefer food to cash. Such food aid can help maintain adequate intakes of protein and energy but is generally not micronutrient rich. That is changing, however, with the inclusion of more diverse foods in the rations provided by the World Food Program as well as supplementary multinutrient foods (Gentilini and Omamo 2011). Three options include local procurement of food aid, which may help small producers and maintain food markets; increased nutrient-density of the food with ready-to-use supplementary or therapeutic foods;[7] and sprinkles, powders that provide multiple nutrients and are mixed with staple food within the home.

School feeding

The 2008–09 crisis and the ongoing one have seen enhanced demand for school feeding programs in low-income countries, despite questions on their cost-effectiveness (Alderman and Bundy 2011). School feeding should be considered as a conditional in-kind transfer to assist low-income households (reducing current poverty) with a complementary benefit of promoting the accumulation of human capital by jointly influencing health and education. It takes two main forms: meals at school and take-home rations. In most middle-income countries, school feeding costs per child represent 10–20 percent of per child costs of basic education, but in some low-income African countries, these feeding costs are as high as the cost of basic education for average students (Bundy et al. 2009).

Programs may yield nutritional benefits for younger siblings of beneficiary children. In Burkina Faso, for example, the weight for age of children aged 12–60 months, whose sister received take-home rations, increased by 0.38 standard deviation. In Uganda younger siblings of beneficiaries of school meals showed improvements in height for age of 0.36 standard deviation, but children in families that took home rations saw no improvement.

School feeding programs can be a vehicle for improved micronutrient status if the foods are fortified, but local procurement issues make fortified foods difficult to obtain. Implementation issues may also affect the overall effectiveness of the program, especially in remote areas where transport and storage costs may be prohibitive for communities. However, school feeding is easy to scale up during a crisis. More evidence is needed on the costs of delivery and sustainability.

Workfare-with-nutrition

Food or cash-for-work programs may provide immediate consumption relief in a crisis. The transfer selection (food or cash) depends on local capacity, market conditions, and cultural acceptability. Some evidence is available from Ethiopia on improved food security and child weight for height from a food-for-work program, but targeting needs to improve. In Indonesia, however, transfer of rice, cooking oil, and legumes had no effect on child growth or on maternal anemia rates (Wodon and Zaman 2008).

A promising new design complements workfare with nutrition interventions. Self-targeting in workfare is relatively effective, because it attracts workers from very poor households where children face a high risk of undernutrition. The approach in Djibouti (and Niger) is to add a nutrition promotion component to the traditional cash-for-work program to leverage the effect of the additional income on the family's nutritional status.[8] In Djibouti (Silva 2010), the workfare component offers community work (for all able-bodied adults) in projects chosen and built by the community and services work (for women only) including collecting, recycling, and transforming plastic bags into pavement blocks. The nutrition component targets vulnerable nonworking members of participating households. It includes monthly community meetings on nutrition-relevant topics, biweekly home visits by a community worker, and distribution of food supplements during the lean season.

Biology and health

There are known evidence-based, effective solutions to undernutrition. Inadequate dietary intake causes weight loss (acute undernutrition), growth faltering (chronic undernutrition), decreased immunity, and increased morbidity and severity of diseases. The Scaling Up Nutrition (SUN) framework (World Bank 2010a; box 2.5) identifies a package of 13 key interventions in three main areas, selected for their efficacy and readiness for scaling up:

- *Behavior change interventions* include. Breastfeeding promotion and support; complementary feeding promotion through counseling and nutrition education (but excluding provision of food), and hand-washing with soap and promotion of hygiene behavior. The majority of these services are delivered through community-based health and nutrition programs.
- *Micronutrient and deworming interventions* include vitamin A supplementation; therapeutic zinc supplements for management of diarrhea; multiple micronutrient powders; deworming; iron-folic acid supplements for pregnant women; iron

BOX 2.5   The global SUN movement

One platform for increased support to nutrition is the global Scale Up Nutrition (SUN) movement. In 2010 more than 100 organizations including governments, civil society, the private sector, research institutions, and the United Nations system committed to work together to fight hunger and undernutrition, developing a Framework to Scale Up Nutrition (launched at the World Bank/International Monetary Fund Spring Meetings in April 2010) and a road map that lays out the operational approach for increased action. SUN is a movement that brings organizations together to support national plans to scale up nutrition. It helps ensure that financial and technical resources are accessible, coordinated, predictable, and ready to go to scale. Twenty-six countries have joined the movement as of early 2010, and the World Bank has agreed to be the donor convenor or co-convenor in 7 of the 26 countries, liaising between the national nutrition focal point and the community donor partners in each of the SUN countries for which it has taken on the coordinating role.

fortification of staple foods; salt iodization; and iodine supplements for pregnant women if iodized salt is not available. These services are delivered through child-health days, community nutrition programs, the primary health care system and market systems (fortification).

- *Therapeutic feeding interventions* include prevention and treatment of moderate undernutrition among children 6–23 months of age, and treatment of severe acute undernutrition with ready-to-use therapeutic foods. These services are delivered through community nutrition programs and the primary health care system.

Community growth monitoring and promotion programs offer a common platform for delivery of multiple services and have been successful in various countries (box 2.6). The community basis allows programs to tackle a wide variety of causes of undernutrition, often with a focus on women and children under age two. The programs have contributed to changing norms about nutritional status and children's growth. Peru has built a local information campaign (RECURSO) to show parents that short stature is a sign of undernutrition and to increase their "demand for good nutrition" (Walker 2008). New tools for measuring, for instance, mid-arm circumference or for visually tracking growth in height are important to raise parental awareness of the dangers of excessive thinness, overweight, and short stature.

Health interventions can also help improve nutrition outcomes through specific services to young children and pregnant and lactating women, including preventing and treating all causes of anemia, promoting good feeding and nutritional care practices, preventing and treating illnesses (especially diarrhea, acute respiratory infections, measles, malaria, and HIV/AIDS), and improving reproductive health and family planning (World Bank 2012 forthcoming).[9] Provision of these services requires that basic health funding be protected during crises, which can be a challenge for countries.

Production and agriculture

There is considerable momentum, including that catalyzed by the global SUN framework, to better link the food security (mainly agriculture) and nutrition security agendas so that countries can benefit from potential synergies. Some SUN interventions have a strong gender component because women face barriers to access inputs and productive assets in many countries, and increasing women's

---

BOX 2.6   Community-based growth promotion programs

Honduras, Jamaica, Madagascar, Nigeria, Senegal, Tanzania, and some states in India use a strategy of community-based growth promotion, which incorporates some of the key Scaling Up Nutrition interventions and strengthens knowledge and capacity at the community level.

Such strategies have proven effective in improving mothers' child-nutrition knowledge, attitudes, and practices; in boosting family demand for health care; and in reducing undernutrition. Successful, large-scale child growth promotion programs in these countries have achieved sharp declines in child malnutrition in the first five years, with a more gradual rate of decline in moderate and mild undernutrition after that. The community basis allows practitioners to address multiple causes of malnutrition, with a focus on women and on children under age two.

Leading interventions include nutrition education or counseling. These interventions often accompany child growth monitoring, offer advice on maternal care services during pregnancy, promote exclusive breastfeeding and appropriate and timely complementary feeding, encourage health and care practices, and make referrals to health centers. Some programs have provided micronutrient supplements for pregnant mothers and children, as well as immunization and related services.

Program experiences highlight the importance of three elements: female community workers as service delivery agents; regular child growth monitoring (weight), paired with counseling and communication with the mother by a well-trained agent who benefits from regular supervision in weighing, recording, and counseling; and well-designed, culturally appropriate, and consistent nutrition education to promote specific nutrition practices. The challenges relate to agent training, support, and motivation; barriers faced by beneficiary mothers in implementing recommended behavioral changes; and high costs of food-supplementation programs for mothers and children.

Source: World Bank 2012 forthcoming.

---

access to human capital is critical to reducing poverty and undernutrition. Changes in agriculture affect health and nutrition through several levers (Hoddinott 2011; World Bank and IFPRI 2008):

- Increased agricultural production may increase household income, which can be used to purchase goods that affect health and nutrition or can be saved in the form of assets, such as improved shelter and access to sanitation, that improve health.
- Changes in agricultural production may result in improvements in household diets, especially through diet diversification and potentially through biofortification of crops (such as vitamin A–rich rice and sweet potatoes).
- Changes in crops or in production processes may make agricultural work more or less physically demanding and may change exposure to pesticides, animal diseases

that can be transmitted to humans, and work-related accidents.
- When returns to agriculture rise, households may increase the labor they devote to agriculture through hiring, decreasing leisure, or increasing child labor.
- Changes in production may result in changes in the intrahousehold resource allocation. Higher earned incomes for women, for example, may affect how money is spent, food is allocated, and the types of assets held, which may improve health and nutrition.

Evidence on these levers is scarce because very few agricultural projects or studies include nutrition in their outcomes, and because agricultural interventions may look less cost-effective than targeted interventions for nutrition alone. The knowledge gap is large, but some studies point to positive impacts of higher income, changes in diet composition, and provision of biofortified foods on nutritional status (Masset et al. 2011).

Strengthening the link between agriculture and nutrition

Some agricultural strategies have strong potential to strengthen the links between agriculture and nutrition. The most promising ones aim to increase the focus on vulnerable groups (like smallholder farmers—particularly women); diversify production (including homestead food production) to increase the availability of legumes, vegetables, and animal-source foods; reduce the impact of waterborne diseases and diseases transmitted from animals; and combine nutrition education with agricultural activities (Pinstrup-Andersen 2010; World Bank 2007; World Bank and IFPRI 2008; World Bank 2012 forthcoming). Dietary diversification is one of the key results for improving diets through own production.

Women as producers are critical to household food and nutrition security in many smallholder economies, especially in Africa. Agriculture interventions need to address the potential negative consequences on household nutrition from increased labor by women. Technology to counteract these effects is often available, but it is rarely accessible to women. In Sub-Saharan Africa, women have less access to fertilizer, labor, and other inputs than men do. But when women secured the same level of inputs as men, they increased their yields for maize, beans, and cowpeas by 22 percent (Quisumbing 1996).

Increasing production of nutrient-dense foods will improve access to diverse diets. Those households producing horticultural crops and raising small animals (poultry, guinea pigs, aquaculture, and the like) will show the greatest improvement in nutritional status. This type of production positively affects the quality of the diet and micronutrient intake. In addition, better preservation of nutrient content or post-harvest fortification can also improve food nutrient content.

A promising range of interventions involves biofortification. The promotion of the orange-fleshed sweet potato (rich in vitamin A) has a direct effect on the vitamin A status of young children and women in Mozambique (Low et al. 2007) and contributes to energy consumption, women's nutritional knowledge and empowerment, and household income. Work by Harvest Plus seeks to strengthen biofortification in iron (effects on anemia), zinc (effects on growth), and vitamin A (night blindness, immune response, and mortality) to address micronutrient deficiencies. Some of the crops that are close to rollout include iron- and zinc-rich pearl millet in India, iron-rich rice in Bangladesh and India, iron-rich wheat in India and Pakistan, iron-rich beans in Rwanda, vitamin A–rich cassava in Nigeria and the Democratic Republic of Congo, and vitamin A–rich maize in Zambia. The 2008 Copenhagen Consensus concluded that biofortification was the fifth most cost-effective intervention to address hunger and undernutrition outside direct nutritional intervention.

Addressing weather variability and seasonal food shortages

Addressing seasonal food shortages through changes in agricultural practices, food preservation, and safety nets can have long-term effects. As noted, the period between conception and two years of age is critical to human development. Because that period covers several agricultural seasons, where seasonal food shortages are typical, children are likely to suffer from some deprivation at some point. With climate change, these seasonal shortages are likely to increase in both frequency and severity. Low-input food preservation technologies (such as solar drying) may increase access to diverse diets for a longer period during the year. Adoption of early or late-season crops, or crops that consume less water, may also help improve diets. Improved water management systems to increase efficient use may improve productivity and also decrease the incidence of waterborne diseases and reduce women's burden of collecting water (Pinstrup-Andersen, Herforth, and Jones 2012 forthcoming). These interventions may be complemented by the provision of social safety nets in the short run. Studies of such interventions in northwest Bangladesh (Khandker, Khaleque, and Samad 2011) show that they are helpful in mitigating seasonal deprivation during the pre-harvest hunger season, especially those administered by NGOs.

Decreasing post-harvest losses

Decreasing post-harvest losses of nutrient-dense foods provides gains to agricultural income and nutrition. Post-harvest loss is especially a challenge for perishable fruits and vegetables (Pinstrup-Andersen, Herforth, and Jones 2012 forthcoming), which have high micronutrient content. Access to markets through investment in roads and post-harvest facilities (storage and basic processing) are key for reducing these losses. Farmers' marketing organizations, offering access to price information for example, are also important.

Targeted subsidies

Governments often use agricultural input subsidies to promote food output, but these subsidies generate much controversy because of their fiscal costs, generally poor targeting, possible undermining of local markets, and lackluster results for rural poverty reduction. Simulations based on the Malawi Agricultural Input Support Program, which provides fertilizer and maize seeds, show that results depend crucially on how the subsidies are financed, on the return on public investments that compete for scarce government funds, and on the size of the productivity gains that smallholders reap from increased application of seeds and fertilizer (Buffie and Atolia 2009). The results are much less favorable when input subsidies crowd out infrastructure investment, which in the long term may enable rural households to diversify their livelihood strategy from staple food production and help them reach food security. Simulations based on comparisons between subsidy programs and social cash transfers in Ghana and Malawi (Taylor and Filipski 2012 forthcoming) show that the cash transfers obtained better outcomes for children's undernutrition.

Women's income and girls' education

Women's education (43 percent) contributed more than food availability (26 percent) to decreases in child undernutrition between 1970 and 1995 (Smith and Haddad 2000). Some of the higher undernutrition rates in South Asia may be related to the lower status of women there. Increased women's income, through access to better jobs as a benefit of the provision of child care in urban poor communities in Guatemala (Ruel et al. 2002), through access to alternative income-generation strategies and credit in Bangladesh, India, and Senegal (World Bank 2011a), and through targeting cash transfers and workfare to women, yield better nutritional outcomes for their children through increased consumption, more diverse diets, and better quality of care.

Increasing women's human capital is one of the most effective ways to reduce poverty and to decrease children's undernutrition. Research in Bangladesh, Ethiopia, Indonesia, and South Africa shows that assets that women bring to marriage play a significant role in how the household makes its decisions. Higher women's assets are associated with a higher share of household spending going to education—especially girls'—and a lower rate of illnesses in girls (Quisumbing and Maluccio 2000; Quisumbing and de la Brière 2000). Because mothers' education is a critical input in the care and nutrition of infants, investments in girls' education will benefit their adult incomes and capabilities—and the welfare of their children.

How much would it cost?

The cost of inaction

Undernutrition causes productivity losses to individuals and GDP losses to countries. In India productivity losses to individuals are estimated at more than 10 percent of lifetime earnings, and GDP loss to undernutrition runs as high as 3–4 percent (World Bank 2009). In Tajikistan undernutrition costs an estimated $41 million annually. Workforce lost to deaths from undernutrition costs the country $12.3 million a year, while productivity lost to stunting, iodine deficiency, childhood anemia, and low birth weight costs $28.6 million.

Nutrition interventions

Horton et al. (2010) put the costs of scaling up the minimum package of the 13 interventions in the SUN package at $11.8 billion a year, of which $1.5 billion is expected to be available from wealthier household resources to cover costs for complementary and fortified foods. The total financing gap is therefore $10.3 billion. Such increases in the resources devoted to nutrition interventions would achieve full coverage of the target population in the 36 countries responsible for 90 percent of the world's stunting. Adding 32 smaller high-burden countries would increase costs by 6 percent. The funds would be raised in two steps.[10]

- Step 1. $5.5 billion a year would be raised, including $1.5 billion for micronutrients and deworming, $2.9 billion for behavioral change, and $1.0 billion to build capacities to start scaling up more complex and targeted food-based programs.
- Step 2. $6.3 billion a year would be raised to scale up complementary and therapeutic feeding in resource-poor environments—$3.6 billion on complementary food to treat and prevent moderate undernutrition and $2.6 billion on treatment of severe acute undernutrition.

The set of interventions and steps will not of course be identical in each country and will reflect the national nutrition issues (seasonal variations, rural/urban distribution, protein-energy shortages, and micronutrient deficiencies), the trade position (importer/exporter), and their administrative capacity—those countries with stronger capacity are likely to move faster to the second step.

No one has conducted a global cost-benefit analysis of nutrition interventions (World Bank 2010a), but individual interventions have consistently shown benefit-cost ratios greater than 2:1 (table 2.1 and figure

TABLE 2.1   The annual per capita cost of various nutrition interventions is very low

Interventions	Annual per capita cost
Breastfeeding promotion	$0.30–4.00
Vitamin A supplements	$0.20
Therapeutic zinc supplements	$0.47 (10 days)
Deworming (school age)	$0.32–0.49
Iron supplement	$10–50
Folate fortification	$0.01
Iron fortification of staples	$0.10–0.12
Salt iodization	$0.05

Source: Horton, Alderman, and Rivera 2008.

2.3). Rates of return for behavioral interventions, such as promotion of breastfeeding, range from 5:1 to 67:1, vitamin A supplementation from 4:1 to 43:1; salt iodization, 30:1, and deworming from 3:1 to 60:1. The newer evidence on long-term benefits of improved nutrition in utero and in the first two years of life may mean that the returns are larger still. New approaches such as multi-micronutrient powders (sprinkles), therapeutic foods, and cash transfers through electronic media also make it easier to implement some of these interventions.

FIGURE 2.3   Benefit-cost ratios of various interventions

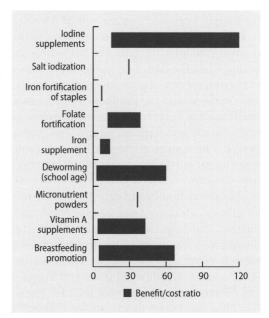

Source: Horton, Alderman, and Rivera 2008.

The global costs of scaling up nutrition interventions are lower than the aid commitments for rural development and agriculture and agro-industries ($14 billion in 2010, see chapter 5) and social safety nets, but they are a big leap from commitments of official development assistance in basic nutrition interventions of $0.3 billion a year during the period 1995–2007. Not all agricultural, health, and social protection interventions are geared to reducing malnutrition, but some interventions in these sectors could bring important nutritional gains, at marginally increased cost. At the country level, spending on safety nets accounted for 1.9 percent of GDP on average before the recent global economic crisis (Grosh et al. 2008; Marzo and Mori 2012). During the crisis, a total of $600 billion was spent on support for safety nets (Zhang, Thelen, and Rao 2010). As noted, some of these interventions can provide platforms to support better nutrition outcomes (World Bank 2012 forthcoming).

Comprehensive, consolidated scaling up of multisectoral nutrition programs implies the need for institutional and policy reforms

Challenges to ramping up investment in nutrition include their multisectoral basis, which requires strengthening coordination between ministries in social sectors, agriculture, rural development, and trade; lack of up-to-date national data on malnutrition, particularly anthropometric data (especially for height),[11] micronutrient adequacy data (blood tests), and behavioral practices such as breastfeeding and hand washing; lack of voice of potential direct beneficiaries (young children and vulnerable pregnant women); and lack of political commitment.

A series of case studies of nutrition policies and programs in countries at differing levels of policy development and program coverage and results have shed light on the process of breaking out of the "low priority cycle."[12] Several factors associated with change were identified, the three most common ones being the coming together of key people who engender confidence about the issue and develop the risk-taking attitude to push for change (the

champions); formation of (broad-based) coalitions and alliances (which often include one or more development partners) that rally behind a common narrative and are able to influence decision makers and decision-making processes; and political "windows of opportunity" that can be seized by the champions and coalitions to push for change (box 2.7).

Sometimes one of the factors may give rise to the emergence of another factor; for example, champions may be able to create political windows of opportunity, or the formation of a coalition may give rise to champions who come forward. In countries, where one or more of the three factors was absent, the push for change generally failed. Conversely, when the three factors became a force for change, it was common to see the development of a shared policy narrative for nutrition, leading to the identification of and focus on selected strategic priorities and the use of strategic communication using data and results to push for institutional development, more resources, or both.

Policy responses and their expected impacts on the nutrition related MDGs

The impact of higher food prices on the MDGs varies across countries and socioeconomic groups (see also chapter 1). Is the country a net exporter or a net importer of the food items for which world prices change? What is the importance (in trade, production and consumption) of the food items for which world prices change? Similarly, the impact of policy responses to world price changes is likely to vary depending on country and policy specifics, including the source of any additional financing that is needed to cover increases in government spending. To explore the impact of expenditure and financing decisions during crises on the MDGs, we extended the MAMS model, a computable general equilibrium model developed at the World Bank for the analysis of country strategies, to cover undernutrition. In this exercise, we assume that food prices double between now and 2015 (and remain

BOX 2.7   Breaking the low-priority cycle: how nutrition can become a public sector priority for Sub-Saharan African governments

In many countries in Africa, the fight against under-nutrition has remained a low government priority for decades, and only recently have some countries begun taking steps to eliminate it.

Political economy factors are important in understanding why, in many countries, nutrition is not recognized as an important priority for human and economic development. Nutrition in many countries is trapped in a "low-priority cycle"—a vicious circle that starts with low demand for nutrition services, followed by a weak response by governments that commit little or no resources and end up with ineffective implementation and poor results, which in turn feed into low demand for nutrition, thus perpetuating its low priority.[a] The accumulation of new scientific evidence on the magnitude of undernutrition and its impact on human and economic development is gradually influencing international donor configurations toward a unified call for scaling up nutrition[b] and the necessary repositioning of nutrition as central to development.[c]

In Sub-Saharan Africa, Senegal is an example of a country that has made significant strides in the fight against undernutrition, where nutrition has broken out of the low-priority cycle, and where the change factors mentioned in the text were prominent. Senegal now has a Multisectoral Forum for the Fight against Malnutrition under the Prime Minister's Office; a national nutritional policy and a National Executive Office that ensures the day-to-day management, coordination, and monitoring of the policy; periodically updated, costed strategic plans for nutrition; multiple programs with multiple stakeholders from all sectors; a budget line currently equivalent to $0.20 per capita per year (compared to $0.03 per capita per year in 2002–06) and projected to grow to $0.65 per capita per year by 2016; donor contributions that average between $0.65–0.70 per capita per year; national program coverage; and, importantly, a reduction in chronic undernutrition that is 16 times above the average reduction in Africa as a whole.

a. See http://go.worldbank.org/NGZM0XHLM1 for published country studies on Benin, Ghana, Madagascar, and Senegal.

b. Scaling Up Nutrition (SUN) is an international movement launched in September 2010; see www.scalingup nutrition.org and box 2.5.

c. World Bank 2006.

constant thereafter) and analyze the implications for two archetype low-income countries. The two archetypes represent a median low-income country along several dimensions; their differences are primarily related to different trade structures, representing medians for net food exporters and net food importers in low-income countries (box 2.8).

At the micro level, there may be strong reasons for policy interventions in both country types. While the aggregate impact of rising food prices is positive for the net exporter, specific household groups may be hurt, especially in the short run. For example, households that are net food purchasers may experience a decline in real incomes, particularly if their incomes are not very responsive to the rise in growth (for example, households that rely on remittances from abroad, for which the domestic purchasing power is undermined by currency appreciation) with potential negative impacts on food and nutrition security. Nevertheless, the need for broader interventions is more evident for the net food importer.

Both the type of intervention and the financing have important implications for the success of nutrition interventions in improving MDG indicators. To illustrate some of the issues involved, we constructed six scenarios, each of which involves a policy response by the net food importer to the rise in the price of food. We then compared their impact on MDG indicators with the scenario of a rise in food prices with no policy adjustment. Four of the scenarios (*sub+tax*, *sub+aid*, *sub+bor*, and *sub+spnd*) involve the introduction of untargeted food subsidies sufficient to keep domestic processed food prices constant through 2025 (as in the baseline).

BOX 2.8   The implications of various spending and financing decisions on the MDGs of a low-income country using MAMS

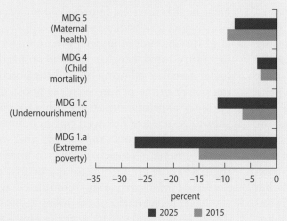

a. Higher export prices for net exporter: Relative changes in MDG indicators compared to baseline of no price increase in 2015 and 2025

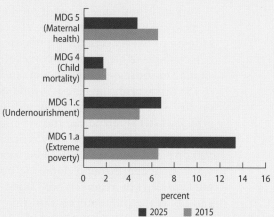

b. Higher import prices for net importer: Relative changes in MDG indicators compared to baseline of no price increase in 2015 and 2025

The left panel of the box figure shows the impact on selected MDG indicators for a low-income country that is in the fortunate position of being a net exporter of food items during a period of rising food prices. We compare two scenarios, one where world food prices are constant through 2025, and a scenario in which world prices gradually are doubled during the period 2012–15, after which they stay at this high level until 2025, the last simulation year. The box figures show the changes in selected MDG indicators in 2015 and 2025 compared with the baseline. The rise in food prices drives increased private demand and government services in response to higher growth, resulting in substantial improvements in the MDG indicators compared with the baseline.

Not surprisingly, as shown in the right panel, the evolution of the same indicators for a less fortunate archetype country—a net importer facing increased prices for its food imports—is the direct opposite. The wide variance in the relative sizes of the gains and the losses for the two archetypes reflects the impact of economic flexibility: both archetypes adjust production and consumption, for the net exporter with the aim of raising exports and for the net importer with the aim of reducing imports.

Source: For more on MAMS, visit www.worldbank.org/mams. For more on the analysis summarized here, see Lofgren (2012 forthcoming).

The source of required additional financing (around 5 percent of GDP) is domestic taxes for *sub+tax*, foreign grant aid for *sub+aid*, domestic borrowing for *sub+bor*, and domestic spending cuts for *sub+spnd* (exempting only transfers to households and spending on agriculture). The last two scenarios (*trn+tax* and *trn+tax2*) impose the same, higher tax rates as *sub+tax* but, instead of subsidizing food, the fiscal space is used for targeted transfers to the bottom halves of the population in rural and urban areas as measured by per-capita income. *Ttn+tax* assumes that this transfer scheme can be handled by the government administration that already is in place, while *trn+tax2* initially imposes additional hiring and other costs amounting to 15–25 percent of the total program cost, declining over time.

The impacts of these policy responses reveal intriguing patterns. The transfer scheme without additional administrative costs (*trn+tax*) achieves the largest reduction in extreme poverty (MDG 1.a) of the six scenarios, followed by transfers with such costs added (*trn+tax2*) and aid-financed food subsidies (*sub+aid*); in this and other respects, the aid-financed food subsidies leave the economy relatively untouched by the import price increase (figure 2.4a). Untargeted food subsidies that are financed domestically are relatively less effective in reducing poverty. Financing via spending cuts (*sub+spnd*) leads to outcomes that are similar to those of tax financing (*sub+tax*); in the real world, the details would depend on the extent to which the spending cuts affect wasteful spending and whether the tax increases distort allocative efficiency or penalize investments. Untargeted food subsidies financed through domestic borrowing (*sub+bor*) do relatively well initially but end up as the only intervention that raises the poverty rate compared with the baseline scenario of no policy action. The primary reason is that increased domestic borrowing reduces domestic private investment and growth in capital stocks and GDP. Initially, this negative impact is relatively minor but over time it becomes important, not unlike undernutrition impacts on a child.

The subsidy schemes are mostly more successful in keeping the rate of undernourishment in check (figure 2.4b), because processed food prices do not increase. Aid financing (*sub+aid*) is preferable, followed by spending (*sub+spnd*) and tax (*sub+tax*) financing, respectively. By 2025 the changes in undernourishment are minor for the remaining scenarios. However, for the two transfer schemes, this limited reduction in undernourishment comes in the context of an increase in real incomes and decisions to reduce food consumption and raise consumption of other items in response to relative price changes. For the case of borrowing-financed subsidies, the main reason that undernourishment does not improve is lower real household incomes.[13] One important dimension to keep in mind is that subsidies in general cover staple foods that are high in calories and low in micronutrients. Even if underweight improves, stunting and micronutrient deficiencies may increase, which has happened in Honduras.

Finally, the impact on MDG 4 (under-five mortality) and MDG 5 (maternal mortality) depends on the impact on growth in real consumption and investment, including private consumption and government health consumption (which translates into government health services; figures 2.4c and 2.4d). Aid-financed subsidies (*sub+aid*) achieve the largest reduction in under-five and maternal mortality rates, because there is no need to make domestic adjustments and the purchasing power of the private sector is boosted by currency appreciation. At the other extreme, subsidies that are financed through government spending cuts (*sub+spnd*) and, to a lesser extent, through domestic borrowing (*sub+bor*) actually raise under-five and maternal mortality rates compared with the baseline scenario of no policy response. These results reflect the negative impact of cuts in government spending on both government health services and the importance of protecting private consumption. The scenarios with tax increases (*sub+tax*, *trn+tax* and *trn+tax2*) have less effect because government services are protected, the decline in private consumption is smaller, or both.

In sum, this analysis suggests that, if administrative costs can be contained, countries should embark on the difficult task of introducing targeted measures, including transfers. If not, untargeted food subsidies may be effective in reducing undernourishment, especially if they are aid financed, because aid has the advantage of making it possible to avoid difficult domestic resource reallocations. However, this does not address stunting and micronutrient deficiencies. In addition, if the subsidies are financed by measures that relatively indiscriminately reduce the resources available for domestic final demands with high payoffs (including private consumption, private investment, and government demand for human development services), then difficult trade-offs emerge and the country may be better off maintaining the status quo. Another important lesson of this analysis is that, to understand the medium- to long-run impact of higher international food prices, it is necessary

FIGURE 2.4   **Impact of policy responses to food import price shock for food net importer**

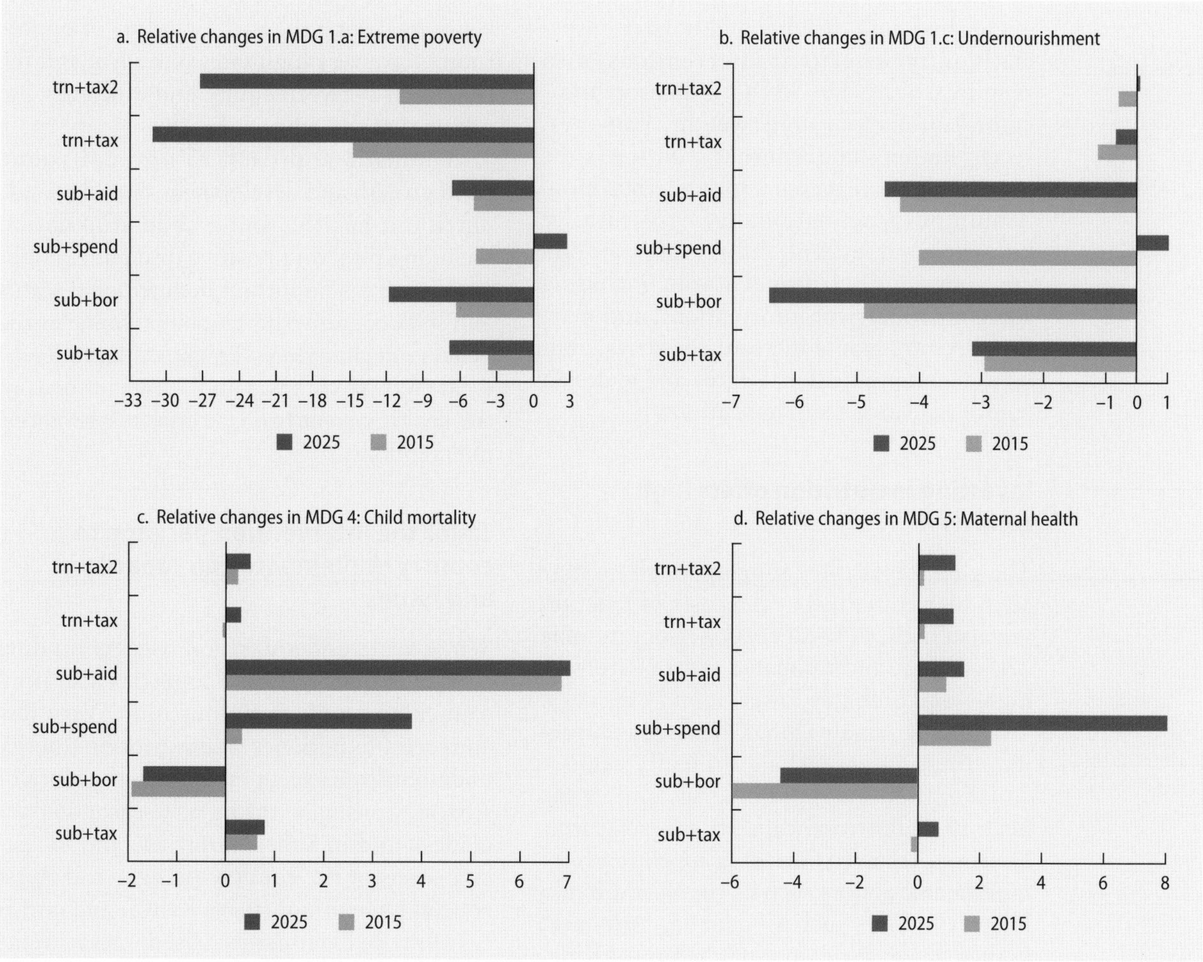

Source: Lofgren 2012 forthcoming.

to consider domestic adjustments and the role of international trade in food for each economy; it would be misleading to assume that food prices change for consumers while everything else remains the same.

Policy Recommendations

Improve the information about nutrition status, practices, and interventions

A basic problem in designing interventions to mitigate the effects of food price hikes is the lack of quality data on basic nutrition indicators and on the effects of both the price rise and some of the interventions to mitigate them. Appropriate responses can be put in place only if countries have a good understanding of who is affected and how. However, few national surveys collect full food consumption data at the household and individual levels with the needed periodicity. Measurement of length or height and weight is difficult, and lack of reliable birth data in some countries makes collecting anthropometric data and computing indexes a challenge. Measurements of micronutrient status often require blood collection, a logistical challenge in many cases, although innovative techniques based on biomarkers may make this information more readily available. Disaggregated data on costs and impacts, especially in multisectoral interventions, also remain scarce. The MDG indicator (indicator 1.b) is child underweight. However, recent findings confirm that stunting is the most appropriate measure for undernutrition. A multipurpose, nationally representative household survey with information on food consumption, nutritional status (including some micronutrient information), and market exposure would increase countries' ability to monitor nutritional status and to design appropriate targeted interventions.

Investing in nutrition offers high returns

The global costs of scaling up nutrition may seem high initially, but the costs of inaction are also high, the unit costs (set out in table 2.1) are low, and estimated returns are very high—and probably lower-bound estimates. Yet funding remains low. One issue is capacity—these interventions typically require collaboration among ministries and in the field. Basic nutrition capacity is also scarce. However, renewed interest is appearing from multilateral donors such as the World Bank; bilateral donors such as Canada, Denmark, France, Japan, Norway, and the United Kingdom; and NGOs such as Save the Children. Increased action may also come through the SUN framework to scale up nutrition (see box 2.5).

Target the period from conception to two years of life

Many interventions have indirect effects on nutrition, but specific interventions for young children and their caregivers and for pregnant and lactating women are crucial, given the importance of that window as a foundation of human capital (see figure 2.1). The earlier evidence about the intensity of physical and sociocognitive development and the negative short-, medium-, and long-term impacts of

undernutrition in utero and in the first two years of life underline this point. Most interventions during the early window of opportunity have very high rates of return, and the trade-offs between equity and efficiency are minimal at this stage.

A holistic approach to optimal young child growth and development should include nutrition, health, young child stimulation including play, and positive discipline. High-quality care is important in nutritional status and sociocognitive development. Some of the behavioral changes will require adaptation to local cultural contexts and a shift of focus of the health system from curative to preventive interventions.

Tailor the intervention package to country implementation capacity and issues

While acute undernutrition triggers funding and relief interventions, countries also need to tackle chronic undernutrition. Very few countries experience acute protein-energy undernutrition except in famines (the Horn of Africa), seasonally (the hungry season in Bangladesh and in Sahel countries), and in specific areas of the country (hunger and thirst zone in Djibouti, northern Kenya, and northeast Brazil). Community-based interventions (see box 2.6) can address acute severe undernutrition, and when food shortages are acute and markets do not function well, food transfers are an important response in the short term but they do not address the prevention of longer-term chronic undernutrition.

"Hidden hunger"—or micronutrient deficiencies—require a different set of interventions. The main micronutrient deficiencies that affect high shares of populations include iron, vitamin A, zinc, and iodine. The package of measures, recommended in SUN, include supplementation to vulnerable groups in high prevalence areas (vitamin A and iron for pregnant women and children, zinc tablets for children with diarrhea), and fortification including iodized salt and fortified flour and sugar. Fortification of staple foods requires collaboration with the private sector. In the future, biofortified crops may contribute to population-level efforts to prevent micronutrient deficiencies. Deworming is also important in settings where women and children have high worm burdens and develop anemia. The ministry of health is commonly the agency responsible for the delivery of deworming, infant and young child feeding programs, and micronutrient supplementation efforts. Community-based programs are frequently the platform for behavior change interventions and nutrition surveillance.

Importers and exporters of food would use slightly different packages to address increases in food prices. However, all countries should build a safety net that can be expanded in a crisis. While general food subsidies are important political tools to maintain food prices at acceptable levels, their fiscal costs and paltry nutritional gains make them less appealing than targeted subsidies or cash transfers to the poor and vulnerable. Computable general equilibrium analysis suggests that, if administrative costs can be contained, countries should embark on the difficult task of introducing targeted measures, including transfers. If not, untargeted aid-financed food subsidies may be effective in reducing undernourishment. However, if untargeted subsidies are financed by measures that reduce the resources for other human development services, then the country may be better off refraining from engaging in untargeted subsidies.

Targeting poor households with young children is one way to improve nutrition outcomes among the groups at highest risks for irreversible negative impacts of undernutrition. A point of entry on nutrition is a comprehensive growth monitoring and promotion program for children, whether at the community level or through the health sector. This program would include information campaigns (such as the one used by RECURSO in Peru) to help mobilize the population and raise awareness about the long-term consequences of undernutrition and the need to shield children and pregnant women from its effects. Box 2.9 describes Haiti's strategy to restore nutrition security after the 2010 earthquake and the first programmatic steps the country contemplates for each priority.

Incorporate nutrition-sensitive approaches in multisectoral interventions

In developing a twin-track approach to nutrition and food security, countries need to weigh the benefits and costs of short-term relief and longer-term investments to raise productivity, especially for smallholder farms, and to work across sectors, especially to link nutrition to health, agriculture, and social protection. A variety of approaches can make interventions in health, agriculture, and social protection—including food aid—more nutrition sensitive (World Bank 2012 forthcoming; see also http://www.securenutrition platform.org/Pages/Home.aspx). Global measures on food trade shape the environment in which decisions are made and have important consequences for national policy options (chapter 4). National markets also matter and need improved functioning (better price information and fewer distortions) and more involvement of the private sector.

Locally, successful implementation will require an alliance of governments with the private sector, NGOs, and communities, especially because an increase in food prices will have disparate impacts depending on markets and production potential. In many cases, behavior and social norms may have to change, with targeted awareness-raising campaigns. Large-scale community nutrition programs have been successful in several low-income countries (see box 2.6). And sometimes NGOs can help expand awareness and coverage. The private sector has a key role in fortification and sometimes in supplementation as well as in improving the availability, accessibility, and affordability of highly nutritious foods.

BOX 2.9   Nutrition security in Haiti after the earthquake of 2010: Priorities and first steps

Nutrition security encompasses access to a nutritious diet, a safe environment, adequate health care, and proper child care practices.

Priorities	First steps
Reduce chronic undernutrition through improved exclusive breastfeeding and complementary feeding practices	Promote behavior change through community education- and household-level outreach
Reduce anemia among pregnant and lactating women and children by providing iron supplements iron and deworming treatments	Provide routine micronutrient supplements (iron, iodine, and vitamin A) to pregnant and lactating women and children under two years
Reduce vitamin A deficiency through supplementation	
Reduce iodine deficiency through supplementation and salt iodization	Reestablish salt iodization
Reduce chronic food insecurity through improved agriculture, investment in agribusinesses, and multisectoral collaboration	Invest in agriculture and agribusiness to increase access to nutrient-rich foods and promote the production of fortified complementary food for children 6–24 months
Improve the coverage of basic health and nutrition services by ensuring proper attention to pregnant and lactating women and children under two years	Invest in basic health services to expand access and quality and include a basic nutrition package for the most vulnerable
Support government capacity and leadership to set, promote and implement nutrition security programs and policies	

Source: World Bank 2010b.

Notes

1. The term *malnutrition* refers to undernutrition (the outcome of insufficient food intake and repeated infectious diseases, including being underweight for one's age, too short for one's age (stunted), dangerously thin for one's height (wasted), and deficient in vitamins and minerals resulting in micronutrient malnutrition) and overnutrition (overweight and obesity). Prevalence of undernourishment refers to the proportion of a population whose dietary energy consumption is less than a predetermined threshold. This threshold is country specific and is measured by the number of kilocalories required to conduct sedentary or light activities.

2. According to the Food and Agriculture Organization, food security is a situation where "all people, at all times, have physical, social and economic access to sufficient, safe and nutritious food to meet their dietary needs and food preferences for an active and healthy life."

3. Morocco's Targeting and Social Protection Strategy (World Bank 2011b) delineates some of the trade-offs and calls for targeting and a different set of interventions to tackle the risks facing the most vulnerable population groups.

4. This section draws heavily on World Bank (2006).

5. This section draws on Alderman (2011a).

6. Gertler (2004) also show results on child health through increased access to preventive health services.

7. Supplementary food contains all the recommended daily allowance of micronutrients along with energy; typically it is a fortified cereal and legume blended flour and is used to address moderate acute malnutrition. Therapeutic food contains all nutrients for children to reverse growth failure and achieve catch-up

(it addresses severe acute and chronic malnutrition). The lipid-rich food is ready-to-eat from its container, requires no water for preparation, is good for 24 months after manufacture and 24 hours after opening.

8. Evaluations are under way in both countries.

9. Birth spacing, adolescent pregnancies when the mother is still growing herself, and sexually transmitted infections all affect fetal growth and infant nutritional status. The longer the interval between birth and the next conception, the more time the mother has to recover nutritionally from her previous birth.

10. In both steps, $0.1 billion is included for rigorous monitoring and evaluation.

11. These data are improving with the implementation of the Living Standards Measurement Study (LSMS) household surveys, as well Demographic and Health Surveys and Multiple Indicator Cluster Surveys in some countries, but sustained funding for regular national-level household surveys is still a challenge. In addition, apart from the LSMS, while the surveys include anthropometric data, they contain very little information on consumption. All these surveys also lack details on household status with respect to food markets (net buyer or seller of the products affected by price hikes).

12. This is based on a comparative study of nutrition policies and programs in Benin, Burkina Faso, Ethiopia, The Gambia, Ghana, Madagascar, Senegal, and Tanzania (led by Marcela Natalicchio and Menno Mulder-Sibanda).

13. In reality, the impact on undernourishment would be more positive than indicated for the two transfer schemes because the inequality of calorie consumption declines; however, in the absence of any data on the distribution of calories per capita other than national Gini coefficients, the analysis could not account for this and assumed instead that this national Gini coefficient did not change.

References

Alderman, H. 2011a. "The Response of Child Nutrition to Changes in Incomes: Linking Biology with Economics." Paper prepared for the CESifo workshop on Malnutrition in South Asia. Venice International University, San Servolo, July 20–21.

Alderman, H., ed. 2011b. *No Small Matter. The Impact of Poverty, Shocks, and Human Capital Investments in Early Childhood Development*. Washington, DC: World Bank.

Alderman, H., and D. Bundy. 2011. "School Feeding Programs and Development: Are We Framing the Question Correctly?" *World Bank Research Observer*.

Alderman, H., B. Ndiaye, S. Linnemayr, A. Ka, C. Rokx, K. Dieng, and M. Mulder-Sibanda. 2008. "Effectiveness of a Community-Based Intervention to Improve Nutrition in Young Children in Senegal: A Difference in Difference Analysis." *Public Health Nutrition*: 12, no. 5: 667–73; doi:10.1017/S1368980008002619.

Armar-Klemesu, M., M. Ruel, D. Maxwell, C. Levin, and S. Morris. 2000. "Poor Maternal Schooling Is the Main Constraint to Good Childcare Practices in Accra." *Journal of Nutrition* 130, no. 6: 1597–607.

Attanasio, O., E. Battistin, E. Fitzsimons, and M. Vera-Hernandez. 2005. "How Effective Are Conditional Cash Transfers? Evidence from Colombia." IFS Briefing Notes BN54, Institute for Fiscal Studies, London.

Baird, S., J. Friedman, and N. Schady. 2011. "Aggregate Income Shocks and Infant Mortality in the Developing World." *Review of Economics and Statistics* 93, no. 3: 847–56.

Baldacci, E., B. Clements, Q. Cui, and S. Gupta. 2004. "Social Spending, Human Capital and Growth in Developing Countries: Implications for Achieving the MDGs." Working paper 04/217, International Monetary Fund, Washington, DC.

Bhalotra, S. 2010. "Fatal Fluctuations? Cyclicality in Infant Mortality in India." *Journal of Development Economics* 93, no. 1: 7–19.

Block, S.A., L. Kiess, P. Webb, S. Kosen, R. Moench-Pfanner, M. Bloem, and C. P. Timmer. 2004. "Macro Shocks and Micro Outcomes: Child Nutrition during Indonesia's Crisis." *Economics and Human Biology* 2, no. 4: 21–44.

Buffie, E. F., and M. Atolia. 2009. "Agricultural Input Subsidies in Malawi. Good, Bad or Hard to Tell?" FAO Commodity and Trade Policy Research Working Paper 28, Food and Agriculture Organization, Rome.

Bundy, D., C. Burbano, M. Grosh, A. Gelli, M. Jukes, and L. Drake. 2009. *Rethinking School Feeding. Social Safety Nets, Child Development and the Education Sector.* Washington, DC: World Bank.

Centers for Disease Control and Prevention. 2011. *Children's Food Environment State Indicator Report.* U.S. Department of Health and Human Services, Washington, DC.

CONEVAL. 2009. Informe de evolución histórica de la situación nutritional de la población y los programas de alimentación, nutrición, y abasto en México.

Cruces, G., P. Gluzmann, and L. F. Lopez Calva. 2010. "Permanent Effects of Economic Crisis on Household Welfare: Evidence and Projections from Argentina's Downturn." Paper prepared for UNDP-RB-LAC project "The Effects of the Economic Crisis on the Well-Being of Households in Latin America and the Caribbean."

Deaton, A. 2010. "Understanding the Mechanisms of Economic Development." Working Paper 15891. National Bureau of Economic Research, Cambridge, MA.

Delisle, H. F. 2008. "Poverty: The Double Burden of Malnutrition in Mothers and the Intergenerational Impact." *Annals of the New York Academy of Science* 1136: 172–84.

Ecker, O., C. Beisinger, and K. Pauw. 2011. "Growth Is Good, but Is Not Enough to Improve Nutrition." Paper prepared for the IFPRI 2020 Conference on Leveraging Agriculture for Improving Nutrition and Health, Delhi, February.

Fernald, L. C. H., P. J. Gertler, and L. M. Neufeld. 2008. "Role of Cash in Conditional Cash Transfer Programmes for Child Health, Growth, and Development: An Analysis of Mexico's Oportunidades." *Lancet* 374, no. 9706: 1997–2005.

Ferreira, F. H. G., A. Fruttero, P. Leite, and L. Lucchetti. 2011. "Rising Food Prices and Household Welfare. Evidence from Brazil in 2008." Policy Research Working Paper 5652, World Bank, Washington, DC.

Ferreira, F. H .G., and N. Schady. 2009. "Aggregate Economic Shocks, Child Schooling and Child Health." *World Bank Research Observer* 24, no. 2: 147–81.

Fiszbein, A., and N. Schady. 2009. *Conditional Cash Transfers: Reducing Present and Future Poverty.* Washington, DC: World Bank.

Friedman, J., S. Y. Hong, and X. Hou. 2011. "The Impact of the Food Price Crisis on Consumption and Caloric Availability in Pakistan: Evidence from Repeated Cross-Sectional and Panel Data." HNP Discussion Paper, World Bank, Washington, DC.

Friedman, J. and N. Schady. 2009. "How Many More Infants Are Likely to Die in Africa as a Result of the Global Financial Crisis?" Policy Research Working Paper 5023, World Bank, Washington, DC.

Friedman, J., and J. Sturdy. 2011. "The Influence of Economic Crisis on Early Childhood Development: A Review of Pathways and Measured Impact." In *No Small Matter: The Impact of Poverty, Shocks and Human Capital Investments in Early Childhood Development,* ed. H. Alderman. Washington, DC, World Bank.

Gentilini, U., and S. W. Omamo. 2011. "Social Protection 2.0: Exploring Issues, Evidence and Debate in a Globalizing World." *Food Policy* 36 (2011): 329–40.

Gertler, P. 2004. "Do Conditional Cash Transfers Improve Child Health? Evidence from PROGRESA's Controlled Randomized Experiment." *American Economic Review* 94, no. 2: 331–36.

Gibson, J., and B. Kim. 2011. "Quality, Quantity, and Nutritional Impacts of Rice Price Changes in Vietnam." Paper presented at the World Bank Poverty and Applied Micro Seminar, September 7.

Grantham-McGregor S., Y. B. Cheung, S. Cueto, P. Glewwe, L. Richter and B. Strupp. 2007. "Development Potential in the First Five Years for Children in Developing Countries." *Lancet* 369, no. 9555: 60–70.

Grosh, M., C. del Ninno, E. Tesliuc, and A. Ouerghi. 2008. *For Protection and Promotion: The Design and Implementation of Effective Safety Nets.* Washington, DC: World Bank.

Haddad, L., H. Alderman, S. Appleton, L. Song, and Y. Yohannes. 2003. "Reducing Child Malnutrition: How Far Does Income Growth Takes Us?" *World Bank Economic Review* 17, no. 1: 107–31.

Headey, D. 2011. "Turning Economic Growth into Nutrition-Sensitive Growth." Paper prepared for the IFPRI 2020 Conference on Leveraging Agriculture for Improving Nutrition and Health, Delhi, February.

Hidrobo, M. 2011. "The Effects of Ecuador's 1998–2000 Economic Crisis on Child Health and Cognitive Development." Department of Agricultural and Resource Economics, University of California at Berkeley. http://ecnr.berkeley.edu/vfs/PPs/Hidrobo-Mel/web/Hidrobo_JMP_1.16.11.pdf.

Hoddinott, J. 2011. "Agriculture, Health and Nutrition: Towards Conceptualizing the Linkages." Paper prepared for the IFPRI 2020 Conference on Leveraging Agriculture for Improving Nutrition and Health, Delhi (February).

Hoddinott, J. and B. Kinsey. 2001. "Child Growth in the Time of Drought." *Oxford Bulletin of Economics and Statistics* 63, no. 4: 409–36 (Also in Alderman 2011b).

Hoddinott, J., J. Maluccio, J. R. Behrman, R. Martorell, P. Melgar, A. R. Quisumbing, M. Ramirez-Zea, A. D. Stein, and K.M. Yount. 2011. "The Consequences of Early Childhood Growth Failure over the Life Course." Discussion Paper 01073, International Food Policy Research Institute, Washington, DC.

Holmes, R., N. Jones, and H. Marsden. 2009. "Gender Vulnerabilities, Food Price Shocks and Social Protection Responses." Background Note, Overseas Development Institute, London (October).

Horton, S., H. Alderman, and J. Rivera. 2008. "Hunger and Malnutrition." Copenhagen Consensus 2008 Challenge Paper. Copenhagen Consensus Center.

Horton, S., M. Shekar, C. McDonald, A. Mahal, and J.K. Brooks. 2010. "Scaling Up Nutrition What Will It Cost?" World Bank, Washington, DC.

IFAD (International Fund for Agricultural Development), WFP (World Food Programme), and FAO (Food and Agriculture Organization). 2011. "The State of Food Insecurity in the World: How Does International Food Price Volatility Affect Domestic Economies and Food Security?" Rome.

Khandker, S. R., M. A. Khaleque, and H. A. Samad. 2011. "Can Social Safety Nets Alleviate Seasonal Deprivation? Evidence from Northwest Bangladesh." Policy Research Working Paper 5865, World Bank, Washington, DC.

Levin, C., M. Ruel, S. Morris, D. Maxwell, M. Armar-Klemesu, and C. Ahiadeke. 1999. "Working Women in an Urban Setting: Traders, Vendors, and Food Security in Accra." *World Development* 27, no. 11: 1977–91.

Lofgren, Hans. 2012 forthcoming. "World Food Prices and Human Development: Policy Simulations for Archetype Low-Income Countries." Background paper for *Global Monitoring Report 2012*, World Bank, Washington, DC.

Low, J. W., M. Arimond, N. Osman, B. Cunguara, F. Zano, and D. Tschirley. 2007. "A Food-Based Approach Introducing Orange-Fleshed Sweet Potatoes Increased Vitamin A Intake and Serum Retinol Concentrations in Young Children in Rural Mozambique." *Journal of Nutrition* 137, no. 5: 1320–27.

Maccini, S., and D. Yang. 2009. "Under the Weather: Health, Schooling, and Economic Consequences of Early-Life Rainfall." *American Economic Review* 99, no. 3: 1006–26.

Macours, K., N. Schady, and R. Vakis. 2008. "Cash Transfers, Behaviorial Changes, and Cognitive Development in Early Childhood: Evidence from a Randomized Experiment." Policy Research Working Paper 4759, World Bank, Washington, DC.

Marzo, F., and H. Mori. 2012. "Crisis Response in Social Protection." Draft background paper for the Social Protection and Labor Strategy, World Bank, Washington, DC.

Masset, E., L. Haddad, A. Cornelius, and J. Isaza Castro. 2011. "A Systematic Review of Agricultural Interventions That Aim to Improve the Nutritional Status of Children." University of London, Social Science Research Unit.

McIntrye, L., N. Glanville, K. Raine, J. Dayle, B. Anderson, and N. Battaglia. 2003. "Do Low-Income Mothers Compromise Their Nutrition to Feed Their Children." *Canadian Medical Association Journal* 168, no. 6: 686–91.

Miller, C. M., M. Tsoka, and K. Reichert. 2011. "The Impact of the Social Cash Transfer Scheme on Food Security in Malawi." *Food Policy* 36, no. 2: 230–38.

Miller, C. M., M. Tsoka, K. Reichert, and A. Hussaini. 2010. "Interrupting the Intergenerational Cycle of Poverty with the Malawi Social Cash Transfer." *Vulnerable Children and Youth Studies* 5, no. 2: 108–21.

Musgrove, P. 1987. "The Economic Crisis and Its Impact on Health and Health Care in Latin America and the Caribbean." *International Journal of Health Care Services* 17, no. 3: 411–41.

Pinstrup-Andersen, P., ed. 2010. *The African Food System and Its Interaction with Human Health and Nutrition*. Ithaca, NY: Cornell University Press.

Pinstrup-Andersen, P., A. Herforth and A. Jones. 2012 forthcoming. "Prioritizing Nutrition in Agriculture and Rural Development Projects: Guiding Principles for Operational Investments."

Pongou, R., J. A. Salomon, and M. Ezzati. 2006. "Health Impacts of Macroeconomic Crises and Policies: Determinants of Variation in Childhood Malnutrition Trends in Cameroon." *International Journal of Epidemiology* 35, no. 3: 648–56.

Quisumbing, A. 1996. "Male-Female Differences in Agricultural Productivity." *World Development* 24: 1579–95.

Quisumbing, A., L. R. Brown, H. S. Feldstein, L. Haddad, and C. Peña. 2000. "Women: The Key to Food Security." International Food Policy Research Institute, Washington, DC.

Quisumbing, A., and B. de la Brière. 2000. "Women's Assets and Intrahousehold Allocation in Bangladesh: Testing Measures of Bargaining Power." FCND Discussion Paper 86, International Food Policy Research Institute, Washington, DC.

Quisumbing, A., and J. Maluccio. 2000. "Intrahousehold Allocation and Gender Relations: New Empirical Evidence from Four Developing Countries." FCND Discussion Paper 84, International Food Policy Research Institute, Washington, DC.

Quisumbing, A., R. Meinzen-Dick, and L. Bassett. 2008. "Helping Women Respond to the Global Food Price Crisis." Policy Brief 7, International Food Policy Research Institute, Washington, DC.

Ramachandran, A. and C. Snehalatha. 2010. "Rising Burden of Obesity in Asia." *Journal of Obesity*. Article ID 868573; doi:10.1155/2010/868573.

Robles, M., and M. Torero. 2010. "Understanding the Impact of High Food Prices in Latin America." *Economia* 10, no. 2: 117–64.

Ruel, M. T. 2000. "Urbanization in Latin America: Constraints and Opportunities for Child Feeding and Care." *Food and Nutrition Bulletin* 21 (1): 12–24.

Ruel, M. T., B. de la Brière, K. Hallman, A. Quisumbing, and N. Coj. 2002. "Does Subsidized Childcare Help Poor Working Women in Urban Areas? Evaluation of a Government-Sponsored Program in Guatemala City." World Bank, Washington, DC.

Ruel, M. T., C. E. Levin, M. Armar-Klemesu, D. G. Maxwell and S. S. Morris. 1999. "Good Care Practices Mitigate the Negative Effects of Poverty and Low Maternal Schooling on Children's Nutritional Status: Evidence from Accra." *World Development* 27 (11): 1993–2009.

Save the Children. 2008. *Saving Children's Lives. Why Equity Matters*. London.

———. 2009. "How the Global Food Crisis is Hurting Children. The Impact of the Food Price Hike on a Rural Community in Northern Bangladesh." London.

———. 2011. "Costing Lives. The Devastating Impact of Rising and Volatile Food Prices." Briefing.

Schady, N. 2002. "The (Positive) Effect of Macroeconomic Crises on the Schooling and Employment Decisions of Children in a Middle-Income Country." Policy Research Working Paper 2762, World Bank, Washington, DC.

Silva, J. 2010. "Djibouti Crisis Response Employment and Human Capital Social Safety Net." Proposal to the Japan Social Development Fund. Middle East and North Africa Human Development Department, World Bank, Washington, DC.

Skoufias, E., S. Tiwari, and H. Zaman. 2011. "Can We Rely on Cash Transfers to Protect Dietary Diversity during Food Crises? Estimates from Indonesia." Policy Research Working Paper 5548, World Bank, Washington, DC.

Smith, L. and L. Haddad. 2000. "Explaining Child Malnutrition in Developing Countries:

A Cross-Country Analysis." Research Report 111, International Food Policy Research Institute, Washington, DC.

Taylor, J. E., and M. Filipski. 2012 forthcoming. "A Simulation Impact Evaluation of Rural Income Transfers in Malawi and Ghana." *Journal of Development Effectiveness.*

Victora, C. G., L. Adair, C. Fall, P. C. Hallal, R. Martorell, L. Richter, and H. S. Sachdev. 2008. "Maternal and Child Undernutrition: Consequences for Adult Health and Human Capital." *Lancet* 371 (9609): 340–57. DOI: 10.106/S0140-6736(07)61692-4.

Victora, C. G., M. De Onis, P. C. Hallal, M. Blossner, and R. Shrimpton. 2010. "Worldwide Timing of Growth Faltering: Revisting Implications for Interventions." *Pediatrics* 125: e473–e480.

Walker, I. 2008. "RECURSO-Perú. La rendición de cuentas a nivél local para mejorar los resultados en servicios básicos." Presentation to International Seminar about Decentralization and Social Inclusion in the Process of Regional Integration. Lima, February 7.

Waters. H., F. Saadah, and M. Pradhan. 2003. "The Impact of the 1997–98 East Asian Economic Crisis on Health and Health Care in Indonesia." *Health Policy and Planning* 18 (2): 172–81.

Wodon, Q., and H. Zaman. 2008. "Rising Food Prices in Sub-Saharan Africa: Poverty Impact and Policy Responses." Policy Research Working Paper 4738, World Bank, Washington, DC.

World Bank. 2005. "Maroc. Réforme du Système de Compensation à la Farine Nationale de Blé Tendre." Report 34454-MA, Washington, DC.

———. 2006. *Repositioning Nutrition as Central to Development. A Strategy for Large-Scale Action.* Washington, DC: World Bank.

———. 2007. *World Development Report 2008. Agriculture for Development.* Washington DC, World Bank.

———. 2009. "République de Djibouti. Options d'Assistance Sociale pour la Lutte Contre la Malnutrition." Note de politique. Région Moyen-Orient et Afrique du Nord, Groupe des Secteurs Sociaux (MNSHD), World Bank, Washington, DC.

———. 2010a. "What Can We Learn from Nutrition Impact Evaluations? Lessons from a Review of Interventions to Reduce Child Malnutrition in Developing Countries." World Bank, Independent Evaluation Group, Washington, DC.

———. 2010b. "Promoting Nutrition Security in Haiti. An Assessment of Pre- and Post-Earthquake Conditions and Recommendations for the Way Forward." World Bank, Washington, DC.

———. 2011a. *World Development Report 2012: Gender Equality and Development.* Washington, DC: World bank.

———. 2011b. "Royaume du Maroc. Note d'Orientation Stratégique sur le Ciblage et la Protection Sociale. " Région Moyen-Orient et Afrique du Nord, Groupe des Secteurs Sociaux (MNSHD), World Bank, Washington, DC.

———. 2012 forthcoming. "Addressing Nutrition through Multi-Sectoral Approaches: Guidance Notes for World Bank Task Team Leaders." World Bank, Washington, DC.

World Bank and IFPRI (International Food Policy and Research Institute). 2008. *From Agriculture to Nutrition: Pathways, Synergies and Outcomes.* Washington, DC.

Yamano, T., H. Alderman, and L. Christiansen. 2003. "Child Growth, Shocks and Food Aid in Rural Ethiopia." Policy Research Working Paper 3128, World Bank, Washington, DC.

Zhang, Y., N. Thelen, and A. Rao. 2010. "Social Protection in Fiscal Stimulus Packages: Some Evidence." A UNDP/ODS Working Paper (March).

Growth and Macroeconomic Adjustment in Developing Countries

Summary and Main Messages

The global recovery shows signs of stalling amid deteriorating financial conditions. Global growth slowed to 3.9 percent in 2011 and is projected to decline further to 3.5 percent in 2012. The strongest slowdown is being felt in advanced economies, but the worsening external environment and some weakening in internal demand is expected to lead to lower growth in emerging and developing countries as well. This outlook is subject to downside risks, such as a much larger and more protracted bank deleveraging in the Euro Area or a hard landing among key emerging market countries. Against these broad developments, food, fuel, and other commodity prices have eased somewhat from their peaks in mid-2011; where high commodity prices had become a concern for broader price stability, this price decline has provided policy makers with greater flexibility to ease monetary policy.

Strengthening the recovery will require sustained policy adjustment at a measured pace that depends heavily on a country's individual circumstances. There are risks in some places of inadequate medium-term fiscal adjustment, and in some of overaggressive short-term fiscal adjustment. In the advanced economies, while fiscal policy consolidation proceeds, monetary policy should continue to support growth as long as unemployment remains high and inflation expectations are anchored. This should be accompanied by steady progress toward repairing and reforming financial systems and by steps to avoid excessively rapid bank deleveraging.

The weaker global economic environment has implications for the emerging and developing countries as they progress toward the Millennium Development Goals (MDGs). Among the low-income countries, despite a solid recovery, a concern is that macroeconomic policy buffers have not been rebuilt to levels before the crisis. Should downside risks such as a sharp global slowdown or another surge in food or fuel prices materialize, these countries will have to confront the situation with weaker buffers than in 2009. In addition to eroded macroeconomic policy buffers, still-high food prices complicate policy making and make progress toward achieving the MDGs more difficult. Accelerated progress toward achieving the MDGs in low-income countries will require adequate and effective international development cooperation and the continued strengthening of policy frameworks in individual countries. Fragile states require special attention.

A weaker global economic environment may impede progress toward the MDGs

Growth slowed in 2011

Global economic growth slowed considerably in 2011 to 3.9 percent, from 5.3 percent in 2010, as the economic recovery continued along two tracks (table 3.1 and map 3.1). In the advanced economies, growth slipped to 1.6 percent, half the rate in 2010 and well below the rate foreseen in the 2011 *Global Monitoring Report (GMR)*, owing to lower than expected growth in the United States and Japan.[1] Modest growth rates were accompanied by relatively high unemployment and low inflation.

In the emerging and developing economies, growth slowed to 6.2 percent, about the level foreseen in the 2011 GMR. Growth in the developing world was led by Asian developing countries, while growth in the Middle East and North Africa was dampened by ongoing political turmoil. Growth in Sub-Saharan Africa continued at around 5 percent, notwithstanding slower export growth to the Euro Area and drought in the Horn of Africa. Despite an overall weaker

global performance, per capita incomes rose in most countries (figure 3.1).

The *World Economic Outlook* of the International Monetary Fund (IMF) foresees a further moderation of global growth in 2012, to 3.5 percent. The Euro Area is expected to be in a recession because of high sovereign borrowing costs, fiscal

FIGURE 3.1   GDP per capita growth

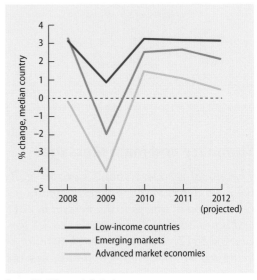

Source: World Economic Outlook.

TABLE 3.1   Global output
Annual percentage change

Region	2008	2009	2010	2011	Projections	
					2012	2013–15
World	**2.8**	**-0.6**	**5.3**	**3.9**	**3.5**	**4.3**
Advanced economies	0.0	-3.6	3.2	1.6	1.4	2.4
Emerging and Developing Countries	6.0	2.8	7.5	6.2	5.7	6.2
Central and Eastern Europe	3.2	-3.6	4.5	5.3	1.9	3.4
Commonwealth of Independent States	5.4	-6.4	4.8	4.9	4.2	4.2
Developing Asia	7.8	7.1	9.7	7.8	7.3	7.9
Middle East and North Africa	4.7	2.7	4.9	3.5	4.2	3.9
Sub-Saharan Africa	5.6	2.8	5.3	5.1	5.4	5.5
Western Hemisphere	4.2	-1.6	6.2	4.5	3.7	4.1
Low-Income Countries[a]	5.9	5.2	6.4	5.5	5.7	6.0
Emerging Market Countries[b]	6.2	2.7	7.7	6.4	5.8	6.3
Fragile States[c]	6.3	3.9	4.3	2.9	5.8	6.3

Source: World Economic Outlook.
a. Low-income countries are those eligible for financial assistance under IMF's Poverty Reduction and Growth Trust, including Zimbabwe.
b. Emerging market countries are emerging and developing countries that are not low-income countries.
c. A subset of emerging and developing countries included in the World Bank's list of fragile and conflict-affected states.

consolidation, and the impact of bank dele-veraging on the real economy. Growth in other advanced economies would slow, in part, because of trade and financial spillover effects from the Euro Area, but is expected to remain positive. In the United States and a few other countries, modest growth momen-tum would be maintained and underlying domestic demand would broadly offset the impact of these spillovers. Overall, advanced economies are projected to grow by just over 1 percent.

In the emerging and developing econo-mies, a weaker and more uncertain external environment, compounded by softer inter-nal demand, is expected to further dampen activity in 2012. Nonetheless, strong growth is expected to continue in developing Asia, in particular China and India. Growth is expected to accelerate in the Middle East and North Africa, led by oil exporters such as Libya, where recovery from the political tur-moil of 2011 is expected; nonetheless, many countries in the region face muted prospects

as political transitions draw out. Countries in central and eastern Europe may be severely affected by trade and financial spillovers from the Euro Area, and recovery there would lag. In contrast, Sub-Saharan African countries should see continued strong growth, except in southern Africa, which is more exposed to weak demand conditions in Europe. Overall, emerging and developing countries are pro-jected to grow by 5.7 percent.

Global current account imbalances remain below those experienced in the run-up to the global financial and economic crisis, and nar-rowed somewhat in 2011 (figure 3.2). Net financial flows to emerging and developing countries, while fairly robust, are also below pre-crisis levels (table 3.2). Average net finan-cial flows were broadly unchanged in 2011 from 2010 and 2009, and the expectation is for similar levels in 2012. Relative to gross domestic product (GDP), low-income coun-tries continue to receive higher net finan-cial flows than do emerging market coun-tries—mainly reflecting significantly higher

FIGURE 3.2   Global current account imbalances

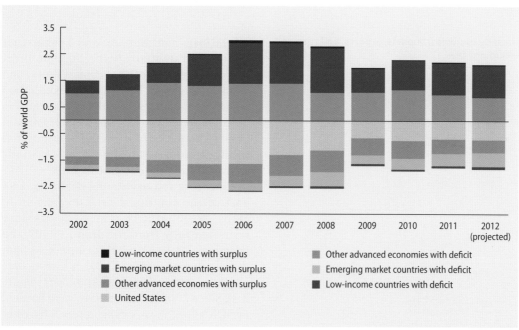

Legend:
- Low-income countries with surplus
- Emerging market countries with surplus
- Other advanced economies with surplus
- United States
- Other advanced economies with deficit
- Emerging market countries with deficit
- Low-income countries with deficit

Source: World Economic Outlook.
Note: The global statistical discrepancy is not shown.

TABLE 3.2 **Net financial flows**
Percent of GDP, equally weighted

Economy	2008	2009	2010	2011	2012 projection
Emerging Market Countries	9.1	7.2	7.4	7.2	6.7
Private capital flows, net	5.4	1.4	2.3	2.9	2.6
Of which: private direct investment	5.1	3.8	3.2	3.3	3.0
private portfolio flows	−1.2	−0.8	0.6	0.3	0.4
Private current transfers	3.5	3.5	3.4	3.3	3.4
Official capital flows and transfers (net)	0.1	2.3	1.7	0.9	0.8
Memorandum item:					
Change in reserve assets (−, accumulation)	−1.7	−2.8	−2.1	−1.5	−1.0
Low-Income Countries	15.4	13.5	13.2	14.7	13.8
Private capital flows, net	4.9	3.2	4.3	4.0	3.2
Of which: private direct investment	6.6	5.3	5.9	6.6	6.3
private portfolio flows	−1.2	−1.2	−1.3	−0.9	−0.9
Private current transfers	5.1	4.7	4.4	4.5	4.5
Official capital flows and transfers (net)	5.4	5.7	4.5	6.2	6.2
Memorandum item:					
Change in reserve assets (−, accumulation)	−2.1	−2.0	−1.7	−2.0	−1.2
Fragile States[a]	15.2	11.4	10.3	19.0	18.8
Private capital flows, net	5.4	2.6	3.9	4.7	5.1
Of which: private direct investment	4.7	3.5	4.8	6.7	6.3
private portfolio flows	−1.0	−1.0	−1.4	−1.6	−1.6
Private current transfers	6.0	6.2	5.9	6.0	5.8
Official capital flows and transfers (net)	3.7	2.6	0.5	8.3	7.9
Memorandum item:					
Change in reserve assets (−, accumulation)	−1.6	−2.0	−2.0	−1.7	−1.9

Source: World Economic Outlook.
a. A subset of emerging and developing countries included in the World Bank's list of fragile and conflict-affected states.

FIGURE 3.3 **Low-income countries: Imports, exports, and current account balance, including FDI**

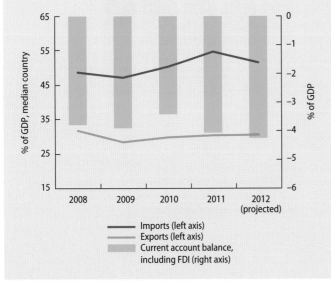

Source: World Economic Outlook.

official loans and grants inflows. Fragile states received substantially higher foreign direct investment (FDI), official capital flows, and official transfers in 2011, and the expectation is that these high levels will be maintained in 2012. While private current transfers remain below the pre-crisis levels, international remittances (in nominal dollars terms) fully recovered from the decrease in 2009.

Emerging and developing countries were part of the continued brisk expansion in global trade in 2011. Their exports have recovered fully from the drop in 2009 and grew by 22 percent in 2011. Current account deficits (net of inward FDI), in low-income countries widened somewhat in 2011 (figure 3.3). Since 2009 (when reserves were boosted by SDR allocations), official reserves have not kept pace with growing trade; however, most emerging and developing countries maintain reserves in excess of three months of

FIGURE 3.4   Official reserves

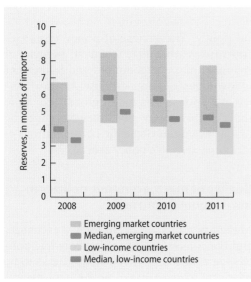

Source: World Economic Outlook.
Note: Bars represent the range between the 25th and 75 percentiles.

imports—one of several measures of reserve adequacy (figure 3.4).

Macroeconomic policies

In advanced economies, ample economic slack and well-anchored inflation expectations continued to provide room for supportive monetary policy. There was much less room for maneuver regarding fiscal policy in 2011, however, given large debt levels and concerns in financial markets over governments' debt sustainability. In emerging and developing countries, increasing prices for food and other commodities through mid-2011 prompted higher headline inflation rates (figure 3.5). Some easing in nonfuel commodity prices since mid-2011 has reduced these pressures, but in many countries commodity price volatility continues to complicate macroeconomic policy making.

After the unprecedented countercyclical fiscal response to the 2009 crisis, emerging and developing countries had begun to reduce fiscal deficits in 2010 and 2011 (albeit rather timidly) (figure 3.6). Although real GDP growth among developing countries in 2011 was similar to that in 2008, fiscal deficits on average (unweighted) remained 2 percentage points of GDP higher than before the crisis.

Among emerging and developing countries that loosened monetary policy in 2011, looser monetary conditions mostly took the form of a nominal depreciation of the currency, rather than a lowering of nominal short-term interest rates (figure 3.7). Against

FIGURE 3.5   Commodity price indexes

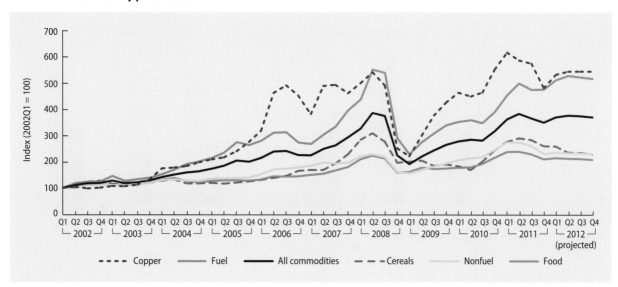

Source: World Economic Outlook.
Note: Indexes are in U.S. dollars. Data for 2012 Q2, Q3, and Q4 are projections.

FIGURE 3.6   **Fiscal deficits in emerging and low-income economies**

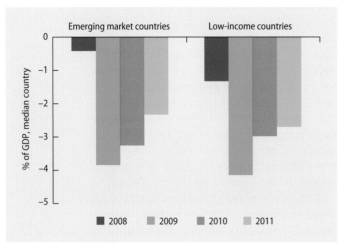

Source: World Economic Outlook.
Note: General government balances as defined in IMF Government Finance Statistics Manual 2001.

this background, monetary aggregates continued to expand broadly in line with the increase in nominal GDP in emerging market countries (figure 3.8).

The direction of macroeconomic policy adjustments varied considerably in 2011. Among the 62 percent of emerging market countries that tightened fiscal policy, more than half complemented that with monetary

tightening (figure 3.9). In contrast, among the 55 percent of low-income countries that tightened monetary policy, half loosened fiscal policy. The variety of policy responses contrasts with 2009, when 90 percent of emerging market economies and 80 percent of low-income countries loosened fiscal policy in response to a major global economic shock. Most policy adjustments seem to be driven by country-specific considerations—including available policy space.

Quality of macroeconomic policies in low-income countries

Monetary policy, access to foreign exchange, and the consistency of macroeconomic policies were judged by IMF country desks to be relatively strong areas of policy implementation in low-income countries in 2011 (figure 3.10). Governance in the public sector, fiscal transparency, and the composition of public spending were assessed as areas of relative weakness. Lower ratings for fiscal policy in 2011 suggest a sense that a return to pre-crisis fiscal positions has progressed too slowly, given the continued strong economic growth rates. Country desks also perceive that the quality of monetary policy slipped

FIGURE 3.7   **Monetary policy loosening in emerging market and low-income countries**

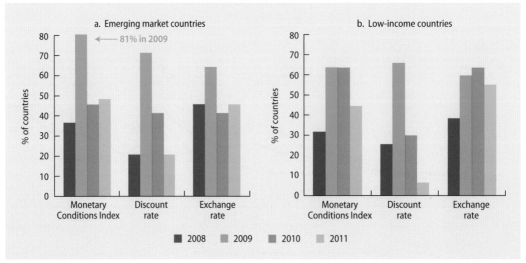

Source: World Economic Outlook.
Note: Monetary policy loosening is based on Monetary Conditions Index (MCI) calculations. The MCI is a linear combination of nominal short-term interest rates and the nominal effective exchange rate (with a one-third weight for the latter).

FIGURE 3.8   Average year-on-year growth in money and the money gap in emerging market countries

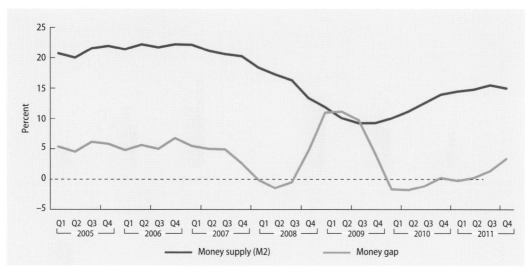

Source: International Financial Statistics.
Note: The money gap is the difference between year-on-year growth rates of M2 and nominal GDP. The sample includes emerging market economies that have data on both for the whole sample period shown.

FIGURE 3.9   Macroeconomic policy mix

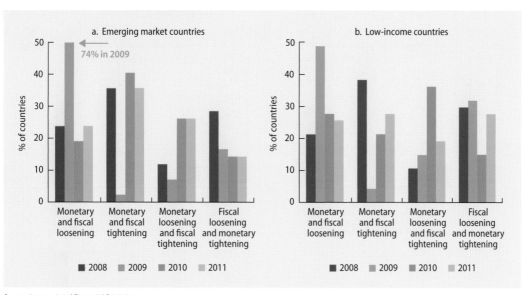

Source: International Financial Statistics.
Note: Fiscal conditions are defined based on annual change in government balance as a percent of GDP in 2008, 2009, 2010, and 2011. Monetary conditions are based on the change in the Monetary Conditions Index; changes are calculated Q4 over Q4, subject to availability (see also figure 3.7's note).

in some countries in 2011; although assessments remained fairly positive overall, more than 10 percent of country desks considered the monetary policy stance unsatisfactory—similar to that of fiscal policy, but noticeably higher than in recent years.

Food price developments and their macroeconomic impact on developing countries

As discussed in chapter 1, food prices have been volatile over the past several years.

FIGURE 3.10   Quality of macroeconomic policies in low-income countries, 2005 and 2009–11

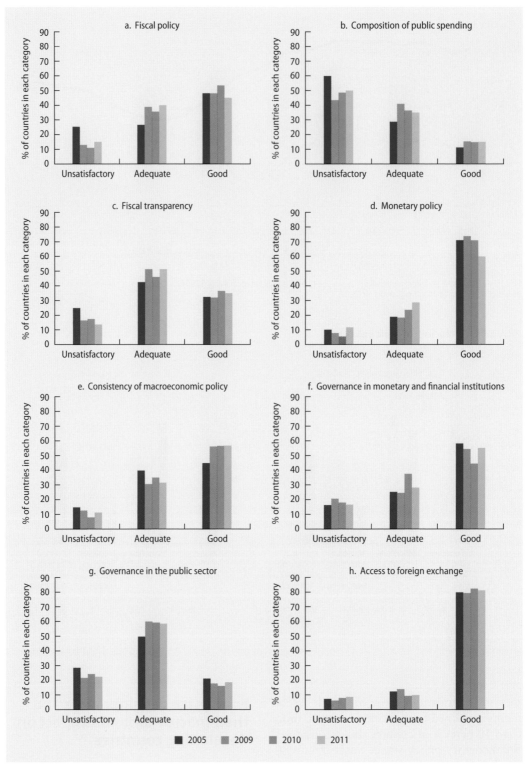

Source: IMF staff estimates.

Global food prices rose by more than 50 percent during 2007 and the first half of 2008, before plummeting by 30 percent in late 2008. By early 2011, however, food prices exceeded the peak level of mid-2008. Although prices have since moderated, average levels in 2011 exceeded those in 2008. While some weakening of prices is projected for 2012 and beyond, the prospects are for relatively high food prices to remain.

An increase in food prices represents a shift of real income away from net-food-importing countries toward net-food-exporting countries. The shift in income takes place through changes in the terms of trade, which affect the purchasing power of domestic firms and households. Countries that are broadly self-sufficient in food will not experience any terms-of-trade losses, but may nonetheless be affected as higher prices trigger a shift of real income from net-food-consuming households to net-food-producing households.

The balance of payments is also directly affected by higher prices for food and other commodities, because changes in terms of trade may trigger payments imbalances. For a typical net-food-importing country, higher

food prices will lead to a widening of the external trade deficit. Initially, the larger deficit may be financed by increased donors' assistance (for example, in the form of food aid) or a drawdown of the central bank's foreign currency reserves. A flexible exchange rate may also work to cushion the balance of payments impact of a higher food import bill (although at higher social costs for vulnerable groups), which over time may be lowered by increases in domestic food production.

While the social implications may be different, and typically less urgent, changes in other commodity prices affect macroeconomic aggregates similarly to changes in food prices. As food, fuel, and other commodity prices often move in tandem, it can be difficult to isolate the effect of food prices alone. The 2007–08 food price surge coincided with even larger increases in fuel prices (figure 3.11). For some oil-exporting developing countries, higher prices for food imports were more than offset by higher prices for oil, while many poorer countries had to confront the challenge of concurrently financing more expensive imports of both food and fuel.

FIGURE 3.11   **Commodity prices and macroeconomic developments, 2007–12**

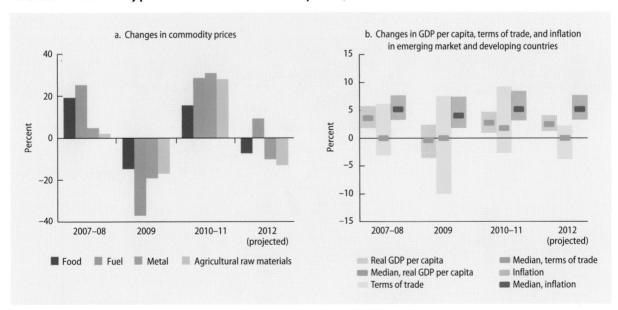

Source: World Economic Outlook
Note: Indexes are in U.S. dollars.

Source: World Economic Outlook
Note: Bars represent range between 25th and 75th percentiles.

MAP 3.1   As global growth slows, growth outcomes across countries converge

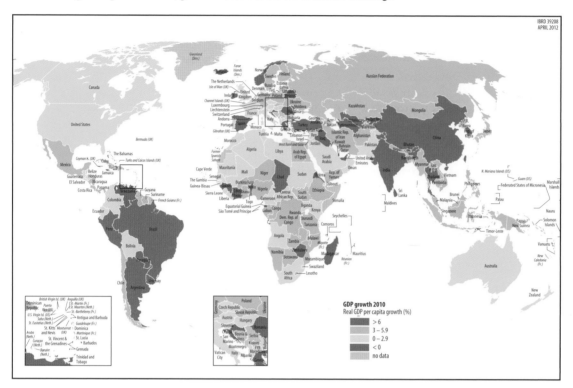

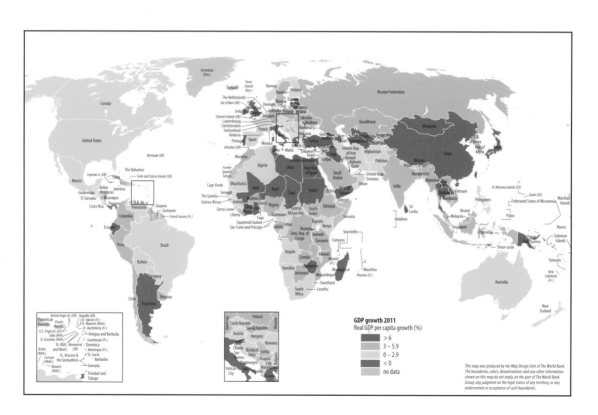

MAP 3.2   With higher commodity prices, few countries are able to maintain price stability

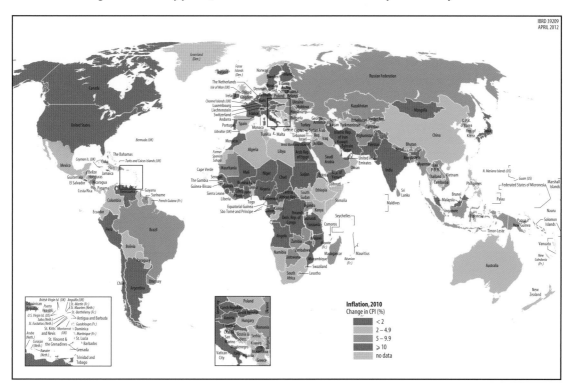

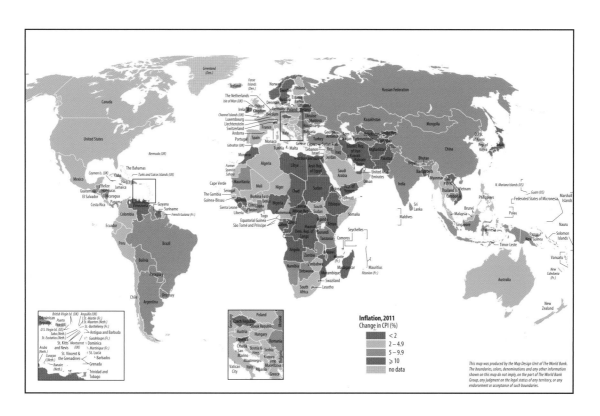

In contrast to the 2007–08 food price shock, the 2010–11 food price shock was part of a broader-based commodity price surge. For many net-food-importing developing countries, the terms-of-trade effects (though not the social implications) of higher food prices were thus mitigated to the extent that these countries export other types of commodities. For example, whereas Sub-Saharan African oil exporters benefited from higher oil prices during both food price shocks, Latin American metal exporters only benefited during the latter episode. In both episodes, robust underlying growth in developing countries cushioned the negative impact on real income in affected countries.

Volatility in food and other commodity prices also has indirect or "second-round" effects. Risk and uncertainty complicate the planning and execution of new investments, dampening investment. These effects can be especially important along the food production chain, where food price volatility makes investment programs less "bankable" and credit more difficult to obtain. Long-lasting price changes may even render part of the country's capital stock prematurely obsolete (a risk in particular when fuel prices rise together with food prices). Such second-order effects affect developing countries' growth prospects both immediately and over the longer run.

To help deal with future negative shocks, developing countries can seek to strengthen their risk management frameworks and can consider securing financial resources on a contingent basis (box 3.1). Contingent financing instruments include commodity-price-hedging instruments, contingent debt instruments (indexed bonds, deferred repayment of loans) or natural disaster insurance (for instance, in the case of a drought). Developing-country demand for these products has been rather limited so far, but examples include oil-importing developing countries that hedge their oil import bill. A few countries have also hedged volatile export proceeds, such as by selling crops in forward markets.

Within a country, higher food prices decrease the real incomes of urban and many rural workers. In poor, vulnerable developing countries, the weight of food in the consumption basket is close to 50 percent and households have limited opportunities to smooth consumption (given their low savings and limited access to credit). Insofar as higher food prices lead poor households to substitute toward less nutritious food items, increased undernourishment may lower health outcomes and cognitive development of children, as discussed in chapter 2. Thus, higher food prices may bring not only immediate economic hardship, but also sustained effects on growth and development. Recent experience underscores how high food prices can also prompt social and political instability that may disrupt economies and weaken economic management, adversely affecting poverty reduction and growth.

As food and fuel prices rose in 2010 and the first half of 2011, consumer prices rose in tandem in many countries (map 3.2). In emerging and developing countries, the median inflation rate rose from 4 percent in 2009 to 6 percent in 2011, but experiences were mixed. In about one-third of all countries, inflation abated over this period, while the share of countries containing inflation in the low, single digits fell from 60 percent in 2009 to 40 percent in 2011. Nonetheless, the share having double-digit inflation remained steady at about 20 percent; in at least some cases, excessively accommodating macroeconomic policies, rather than cost-push pressures, may have been responsible.

Still, high food and fuel prices noticeably affected inflation levels in several low-income countries. In Burundi inflation more than tripled from 4½ percent in 2009 to 15 percent in 2011 as the monetary authorities sought to contain the second-round effects of the imported inflation. Inflation doubled in Bangladesh from 5½ percent to 11 percent over the same period. Other examples of sharp increases in consumer prices during this period that were associated with international food and fuel prices include the Kyrgyz Republic (from 7 to 17 percent), Maldives (from 4 to 12 percent), and Mozambique (from 3 to 11 percent).

The food price shocks of 2007–08 and 2010–11 constituted adverse shocks in many

---

BOX 3.1　Dealing with shocks: Risk management and contingent financing instruments

Adverse external shocks, even when temporary, can have prolonged negative effects on income and poverty in developing countries. Natural disasters or sharp swings in commodity prices or export volumes, for example, can disrupt growth and affect the fiscal and balance of payments positions, which in turn may threaten core public spending on health, education, and infrastructure.

To mitigate the impact of these shocks, countries require an appropriate risk management framework and access to a range of risk management tools. Since the types of shocks and the degree of risk are specific to each economy, a risk management framework begins by assessing the country's principal fiscal risks and debt sustainability vulnerabilities, including by analyzing fiscal flows, the government balance sheet, and contingent liabilities. The World Bank assists developing countries in creating effective risk management frameworks; the International Monetary Fund also provides support in key areas such as fiscal risks and asset and liability management.

Risk management tools are of three broad types: self-insurance; ex post financing arranged after a shock hits; and ex ante financing arranged before a shock hits. Countries self-insure against shocks by building up official reserves and other macroeconomic policy buffers. But there are limitations. For example, public investment and other development needs imply a high opportunity cost to holding excessively large reserves.

External finance can complement self-insurance. Particularly for low-income countries, which often require grants or low-interest loans, an effective architecture for the financing of shocks should provide predictability while still delivering scarce concessional resources in amounts tailored to countries' needs stemming from a shock. Financing arranged after a shock can be better tailored and can limit moral hazard, but its volume and timing is not assured in advance; complementing it with the possibility of ex ante support could give greater confidence to policymakers in low-income countries that at least part of their needs would be met promptly in the face of shocks.

This ex ante support can be provided by contingent financing instruments such as insurance, market hedging, contingent credit lines, and contingent debt instruments. The historically low use of contingent financing instruments by low-income countries partly reflects factors such as affordability, political economy concerns, and technical capacity. The international community can help in addressing some of these constraints to the use of contingent financing instruments.

Source: IMF 2011c.

---

emerging and developing countries. These shocks, however, were partly offset by relatively buoyant economic conditions. As global growth faltered in 2009, the developing world was negatively affected, but at the same time lower food prices provided a respite. A similar scenario of moderating growth in the context of lower food prices could occur in 2012.

Managing macroeconomic risks in developing countries

Risks to the baseline outlook

There are important downside risks to the global baseline outlook explored here.

Perhaps the most immediate is the possibility of a larger, more protracted bank deleveraging in the Euro Area. Tightening credit would deepen the recession and further strain fiscal positions, with additional spillovers. Deleveraging could also affect emerging and developing countries more directly: Euro Area banks account for large shares of global trade finance, an area where the impact of deleveraging was already evident by late 2011. Another key risk is that medium-term fiscal consolidation plans and rising medium-term public debt levels could leave Japan and the United States vulnerable in the event of turmoil in global bond and currency markets. In key emerging market countries, where growth has benefited from buoyant credit markets and asset price increases, a hard landing that triggers a loss of confidence and an unwinding of credit and real estate markets could slow growth significantly.

Food and other commodity price projections are also subject to risks. For example, increased geopolitical tensions may push up fuel prices with knock-on effects on other commodity prices (through higher transportation and other production costs). Higher fuel prices may also lead to a further expansion of biofuel production at the expense of food production, thus placing further pressures on food prices. If global growth is higher than expected, demand pressures could also lead to higher commodity prices, at least in the short run. Although food prices are expected to continue to ease in the period ahead, a broad range of alternative scenarios could well lead to a retesting of the peak food prices of 2008 and mid-2011.

The low-income countries may be particularly vulnerable to these risks

Most low-income countries recovered swiftly from the global crisis and growth has been strong since early 2010, helped by past macroeconomic and structural reforms that had enhanced the resilience of their economies. Nonetheless, with their macroeconomic buffers still well below pre-crisis levels, most low-income countries are now less prepared to cope with further external shocks. As analyzed by IMF staff in the fall of 2011, adverse shocks to global growth and commodity prices could thus have severe economic and social consequences.[2] At the peak of the global crisis in 2009, many low-income countries used strong pre-crisis macroeconomic buffers to pursue countercyclical fiscal responses: despite falling revenues, they maintained and often even increased spending. While growth recovered swiftly from the global crisis, most low-income countries have since made little progress in rebuilding those buffers. Fiscal adjustment began in 2010 as revenues rebounded, but has since halted—in part because of measures taken in response to the commodity price shock of early 2011.

Current account deficits (net of FDI) have widened, especially for net oil importers. And reserve coverage has declined since the 2009 IMF special drawing rights allocation, in particular for many low-income countries with pegged exchange rates. Consequently, most low-income countries are now less prepared to cope with further external shocks than they were in 2008 (figure 3.12). In the event that downside risks materialize, for most low-income countries the scope for fiscal stimulus would be more limited than in 2009, given weaker fiscal buffers and constrained aid envelopes.

To provide a more structured assessment of these vulnerabilities, an analytical framework was used to simulate two (mutually exclusive) tail-risk scenarios for all low-income countries (figure 3.13):

- A *sharp downturn in global growth scenario*, in which shocks to financial conditions in advanced economies reduce global growth by 1.3 percentage points in the first year and by 1.6 percentage points in the second year, relative to the World Economic Outlook baseline.
- A *spike in global commodity prices scenario*, involving surges in prices for food (25 percent in the first year and 31 percent in the second year), fuel (21 percent and 48 percent), and metals (21 percent and 36 percent), relative to the World Economic Outlook baseline.

The adverse global growth shock is estimated to cut about 1 percentage point off low-income country growth in each of two years, because these countries are negatively affected through channels such as global export demand, commodity prices, remittances, and FDI. The severity of the impact would vary, with more than a quarter of low-income countries experiencing a growth slowdown exceeding 2 percentage points. A downturn in global growth would severely erode external and fiscal buffers, causing fiscal deficits to increase (by about 1 percentage point of GDP for the median low-income country) and, for most low-income countries,

FIGURE 3.12   Selected macroeconomic indicators for low-income countries, 2007–12

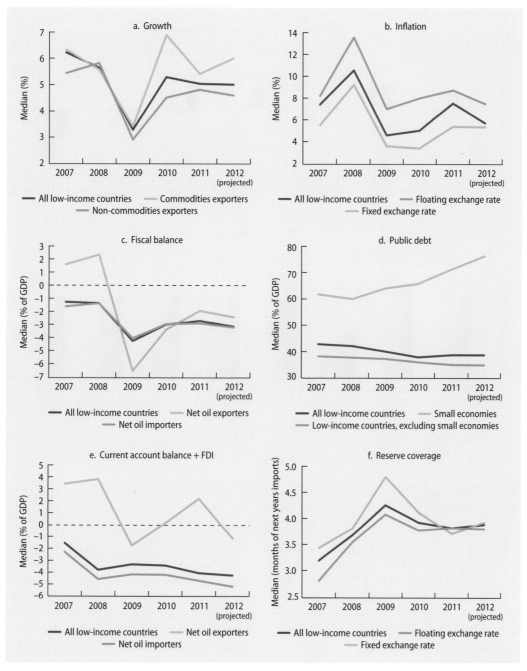

Source: World Economic Outlook.

current account deficits to widen and official reserve coverage to fall.

The *price spike scenario* recognizes that commodity price shocks tend to create winners and losers both within and across countries, depending on the terms of trade, sectoral employment, and consumption patterns. However, its repercussions on inflation, poverty, and social pressures would be felt more symmetrically, because of high shares

FIGURE 3.13   **Tail-risk scenarios for low-income countries**

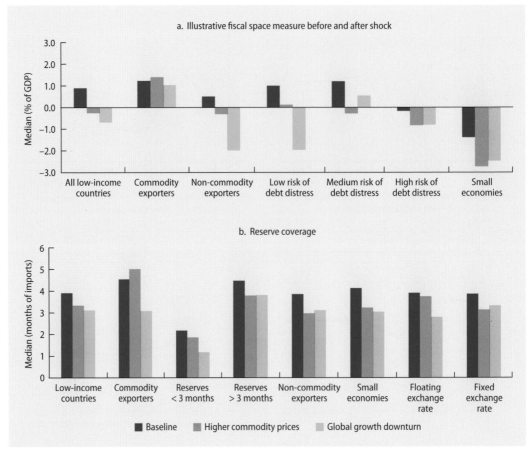

Source: IMF 2011b.
Note: The illustrative fiscal space measure (top panel) is calculated as the difference between the baseline primary balance and the constant primary balance that is needed to achieve a target public debt-to-GDP ratio of 40 percent in 2030. The bottom panel shows a simulation of the reserve coverage ratio after an increase in global food, metals (except gold and uranium), and fuel prices (by 31, 36, and 48 percent respectively) and a slowdown in global growth (by 1.6 percentage points) relative to the World Economic Outlook baseline.

of food in the consumption baskets of low-income countries. While the growth impact of this scenario would likely be modest, inflation could more than double, assuming that the pass-through from global to domestic prices follows historical patterns and that any monetary policy response is mild.

The external impact of a commodity price spike would differ significantly across low-income countries depending on their trade structure. A large majority would be adversely affected, however, with the median trade balance deteriorating by almost 3 percent of GDP. For commodity exporters, a

negative median impact from food and fuel prices would be more than offset by the gain from higher prices of other commodities. About one-fifth of low-income countries would stand to gain from higher prices. Among those hurt by the shock, about half would have adequate international reserves to absorb the shock and the others would face additional financing needs.

In many low-income countries, increased global commodity prices would put pressure on fiscal positions, assuming that countries maintain existing policies (such as fuel subsidies) and that they reintroduce transfers and subsidies similar to those used in 2007–08. A sharp increase in commodity prices that was sustained for long periods would also worsen debt dynamics in a number of low-income countries with existing debt vulnerabilities.

Policy responses in the event of adverse external developments

A key policy challenge for many low-income countries is to build resilience while supporting economic development. This requires balancing pressing spending needs, including public investment and social protection, against the rebuilding of macroeconomic buffers to prepare for future shocks.

Many low-income countries could still benefit from a further strengthening of their fiscal buffers. Scenario analysis indicates that a large number of low-income countries can only partially absorb large tail-risk shocks. This group could consider a mix of gradual fiscal adjustment combined with realignment of priorities, for example by shifting spending in favor of investment and social programs, and building a stronger revenue base. Those countries that already have no fiscal space under the baseline would have limited room for maneuver in the event of a tail-risk shock. For this group, rebuilding fiscal buffers and strong concessional support from development partners will be particularly important. Some low-income countries already have adequate fiscal buffers, and may even be able to expand their fiscal deficits in the baseline, for instance to step up critical spending, without compromising their ability to absorb large shocks.

While many low-income countries have built up sufficient reserves to absorb the impact of either shock fully without the need for adjustment (and import compression), others would benefit from building additional reserve buffers. These buffers could be achieved through a mix of monetary and fiscal tightening, combined with greater exchange rate flexibility where appropriate. A quarter of these countries already have import coverage of less than three months

under the baseline. For this group, rebuilding external buffers should be a high priority. These countries are most in need of help from the international community.

To reduce their exposure or create space to prepare for future shocks, low-income countries can also take steps ex ante. Besides building policy buffers during good times, they can, for example, make their budgets more structurally robust (IMF 2011d); put in place more flexible and robust social safety net systems; pursue reforms to encourage domestic savings and deepen their financial sectors; and explore policies to encourage greater diversification in an economy's production and exports. A specific example in this regard would be to lower domestic fuel subsidies. This step would directly strengthen the fiscal buffer, while also giving the private sector incentives to pursue a more rational use of energy. Another example would be to lower import tariffs—at a pace that acknowledges the potential revenue implications—to better align domestic and international prices of traded commodities.

Macroeconomic policies in the event of a sharp global downturn

The appropriate macroeconomic policy response to a sharp global downturn would depend in part on available policy buffers. During the global downturn in 2009, low-income countries with more comfortable buffers were able to mount a strong countercyclical fiscal response that cushioned the impact on spending and growth. In the event of another sharp downturn, the scope for fiscal stimulus would be more limited for most low-income countries, given weaker fiscal buffers and constrained aid envelopes, but those with sufficient fiscal room should aim to protect spending. For countries lacking fiscal room, key challenges will be to limit the decline in domestic revenue to the extent possible through strengthening tax and customs administration, and to prioritize spending. If fiscal space allows, low-income countries should seek to soften the economic and social impact of a global downturn by preserving—and where feasible increasing—real fiscal spending in priority areas.

In the event of another global downturn and related softening in commodity prices, more active monetary easing may be appropriate in low-income countries with moderate inflation. Greater exchange rate flexibility could also help to weather another downturn, and would be particularly important for those countries with low reserve cushions.

Macroeconomic policies in the event of global commodity price spikes

Global commodity price spikes present low-income countries with difficult tradeoffs between price stability, external goals, and social objectives. A pragmatic response could include targeted measures to protect the poor and a monetary policy response that may largely accommodate the first-round impact on inflation, although those countries with limited reserves may need to tighten policies in support of external and price stability. The scope to use tax and expenditure measures to mitigate the social impact of higher commodity prices depends considerably on the country-specific fiscal space.

The illustrative tail-risk scenario indicates that many low-income countries appear to have adequate fiscal space to absorb the effect of a large, but temporary, global commodity price shock. By contrast, those lacking fiscal space even under the baseline would need to adjust over the medium term to preserve fiscal sustainability after such a shock. A "first-best" policy response to global price shocks would consist of fully passing on price increases while relying on an effective, well-targeted social safety net—in combination, these measures would ensure fiscal affordability and avoid economic distortions, while protecting the most vulnerable. However, institutional capacity and political constraints often make the first-best infeasible, particularly in the shorter term. These constraints may imply a need to resort to pragmatic policy responses—a challenge then being to make the measures as cost-effective and targeted as possible. A number of "second-best" policy approaches have

been used—some effectively, and others less so (box 3.2).

The appropriate monetary policy response to a food price shock depends on the inflation outlook, the share of food prices in household consumption baskets, the pass-through from food prices to other prices, and the country's external balance, debt, and reserves situation. When inflation is at low to moderate levels, the standard monetary policy advice is to accommodate the direct impact of the food price shock, while guarding against any second-round effects. (For food importers, adjustment will often require some degree of exchange rate depreciation, amplifying the inflationary impact.) This allows the monetary authorities to avoid an undue policy tightening that would exacerbate the impact of the price shock on output, while preventing a persistent effect on inflation and inflation expectations (box 3.3). However, food-importing (and other commodity-importing) low-income countries with high inflation or weak external buffers such as high current account deficits, low reserves, or vulnerable debt positions) may require policy tightening.[3] Striking the proper balance can be particularly complex for low-income countries, where products that exhibit considerable price volatility, such as food and fuel, may constitute half of the consumption basket (figure 3.14). On the other hand, because wage indexation and other contract mechanisms that foster inflation inertia are less prevalent in many low-income countries, a temporary surge in commodity prices will have milder second round effects on inflation.

A note on fragile states

Fragile states are characterized by weak public institutions, lack of timely and reliable statistics on the basis of which policies can be formulated, skills shortages, slow rates of GDP growth, and greater macroeconomic instability. Peace- and state-building takes priority over formulating and implementing consistent medium-term macroeconomic policy frameworks. Conflict and other major shocks not only bring great hardship but also

BOX 3.2   Fiscal policy responses to food price shocks

In designing policies to respond to food price and related shocks, national authorities consider the effectiveness of various tax and expenditure policies and the fiscal space available to implement these policies without endangering macroeconomic objectives. The scope for mitigating the impact of higher food prices depends considerably on earlier policies and how those have affected the country's fiscal and debt positions. The appropriate fiscal response also depends on the nature of shock and its expected duration.

Even for countries with ample fiscal space that face a spike in food prices, measures aimed at limiting the price increase for all consumers, such as a general price subsidy, are typically not optimal. First, by providing relief to the general population, large shares of the cost of these schemes are incurred in subsidizing consumers that may not require assistance. Second, because broad-based subsidies are more expensive, if the shock persists their cost becomes a greater concern very quickly. Third, political economy considerations can make it difficult to eliminate price subsidies once they are in place. Finally, subsidies create a substantial wedge between world market and domestic prices; incentives for smuggling could lead to the budget subsidizing consumption in neighboring countries.

Developing countries' experiences dealing with recent years' high and volatile oil prices are illustrative in this regard (Granado et al). After oil prices began to rise at the end of 2003, most developing countries limited the full pass-through of international prices to domestic consumers (the median pass-through was lowest in the Middle East and highest in Africa). When oil prices did not subsequently reverse, the cost of maintaining the subsidies mounted and by mid-2008 reached about 1 percent of GDP in affected countries, with most of the associated benefits on household welfare accruing to the better off segments of the populations. Pass-through of international to domestic food prices varies across regions and countries; as discussed in chapter 1, in countries open to trade the pass-through is faster and relatively larger.

While a well-targeted social safety net aimed at the most vulnerable households is preferable to general price subsidies, such safety nets are difficult to design and implement. Until they can be put in place, in some cases policy makers may subsidize particular products predominantly consumed by the poor (such as coarse grains) while recognizing that some nonpoor households may also benefit from the scheme.

Export taxes and restrictions have also been used in an attempt to dampen domestic price increases, but these have considerable drawbacks, including exacerbating the volatility of global prices (chapter 4). Reductions in import tariffs—if only temporary—carry similar drawbacks. Measures to address supply constraints such as agricultural input subsidies—may have a role if implemented within a broader strategy focused on increasing agricultural productivity. However, the experience with input subsidies is mixed (chapter 2).

During both the 2007–08 and 2010–11 food price shocks, countries implemented a broad variety of measures to counteract the effects of higher international prices. Examples of targeted measures include the provision of food vouchers to the lowest quintile households in the two largest urban areas in Burkina Faso, an expansion of school feeding programs in Sierra Leone, and a conditional cash transfer program targeting orphans and vulnerable children in Kenya. Broader across the board measures included a suspension of customs duties on rice, wheat, and powdered milk in Senegal and a suspension of taxes on food products and the introduction of fuel subsidies in Burkina Faso. In Guinea, a reduction in retail prices on fuel turned out to be very costly and spurred illegal reexports to neighboring countries. Vietnam temporarily banned rice exports for a few months until it was clear that the new harvest was sufficiently large. In the meantime, world market prices had started to fall rapidly and Vietnamese exporters experienced larger drops in their earnings than did their Thai counterparts.

During the recent commodity price shock, most low-income countries adopted countervailing fiscal measures to mitigate the impact of higher food and fuel prices. In several cases, the fiscal costs exceeded the measures introduced during the 2007–08 episode. An often-used measure was fuel subsidies. These subsidies helped lower transportation costs, and thus indirectly food prices. The median (annual) fiscal cost is estimated to exceed 1 percent of GDP for those countries adopting the measures. Most fuel or food subsidies were universal, and few were explicitly targeted to the poor. While these fiscal measures helped address urgent economic and social concerns, they also prevented low income countries from making further progress toward restoring their fiscal deficits to levels prevailing before the 2009 crisis.

BOX 3.3 Food price volatility and monetary policy

Recent research done for the International Monetary Fund's World Economic Outlook suggests that central banks faced with high and volatile food prices set and communicate monetary policy based on developments in underlying inflation (IMF 2011e).

This finding hinges on the observation that, when food prices are volatile and the share of food in the consumption basket is high, it can be very difficult to control headline inflation. Food price shocks often stem from weather disruptions and other shocks that are generally temporary and outside the control of the central bank. Consequently, when a food price shock hits, a central bank targeting headline inflation will be faced with either a loss of credibility if it accommodates the shock, or collateral economic volatility if it attempts to dampen the inflationary effects of the shock. Conversely, if a central bank has established and communicated a clear focus on underlying inflation that is embedded in people's expectations, it can successfully accommodate the first-round effects without undermining credibility or risking higher future inflation. A striking consequence is that a central bank can achieve lower headline inflation and output volatility than if it had focused on headline inflation. The key channel for this result is the preservation of the central bank's credibility and the anchoring of inflation expectations when food price shocks hit.

While an emphasis on underlying inflation can deliver superior outcomes, there are challenges in establishing such a regime. A common objection to the use of underlying inflation targets is that they do not necessarily reflect the day-to-day prices faced by consumers. However, even headline inflation is not an accurate measure of the prices faced by any given consumer. For example, consumption patterns of households with many children will be very different from those made up of young adults, and neither consumption pattern will match the basket used for the headline inflation measure. Furthermore, underlying inflation measures are generally constructed so that over the medium run, if not the short run, they show the same average level of inflation as headline inflation. The research argues that the central bank, thus, has some choice over the target used, a finding that is supported by the successful experiences of inflation-targeting central banks that established their regimes around underlying inflation measures through the use of sustained and ultimately successful communications strategies.

Volatile food prices present a significant challenge to central banks trying to control inflation. This challenge is magnified in countries with high shares of food in their consumption baskets seeking to establish a credible policy regime. The research suggests that these central banks would do better to target what they can hit (that is underlying inflation) than valiantly trying to control headline inflation in the face of food price shocks that are outside their control.

FIGURE 3.14 **Composition of the Consumer Price Index basket in low-income and OECD countries**

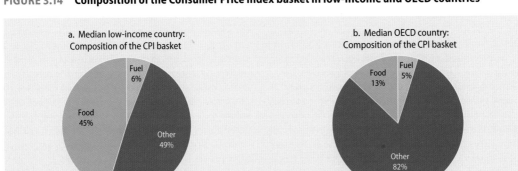

a. Median low-income country:
Composition of the CPI basket

Fuel 6%
Food 45%
Other 49%

b. Median OECD country:
Composition of the CPI basket

Food 13%
Fuel 5%
Other 82%

Source: World Economic Outlook.

set back years of investment in public institutions and public infrastructure, perpetuating a cycle of underdevelopment.

In the face of a slowdown in global growth, the structural problem of unemployment, particularly among the young, would become starker. In the face of food price shocks, fragile states lack many of the policy options available to other developing economies. Because available policy space is strictly limited, these countries often turn to the international community for assistance. To engage most effectively, international organizations and development partners are increasingly recognizing the limited capacity and large financing needs of fragile states, and developing longer-term strategies to benefit them.

Notes

1. The classification of countries is the one used in the IMF's *World Economic Outlook*. Emerging and developing countries are those countries that are not designated as advanced countries. Countries that are eligible for financial assistance under the IMF's Poverty Reduction and Growth Trust constitute a subset of emerging and developing countries; these countries are denoted low-income countries although eligibility is based on other considerations in addition to income levels. Emerging and developing countries that are not eligible for financial assistance under the Poverty Reduction and Growth Trust are designated as emerging market countries.

2. This section draws from IMF 2011b.

3. Food and other commodity exporters should generally rely on exchange rate appreciation to mitigate inflation pressures from a food price spike.

References

Arezki, Rabah, and Markus Bruckner. 2011. "Food Prices and Political Instability." IMF Working Paper 11/62, International Monetary Fund, Washington, DC.

Granado, Del, Javier Arze, David Coady, and Robert Gillingham. 2010. "The Unequal Benefits of Fuel Subsidies: A Review of Evidence for Developing Countries." IMF Working Paper 10/202, International Monetary Fund, Washington, DC.

G-20 (Group of 20). 2011a. "2011 Report of the Development Working Group." October.

G-20. 2011b. "Report of the G-20 Study Group on Commodities." July.

IMF (International Monetary Fund). 2001. *Government Finance Statistics Manual 2001*. Washington, DC.

———. 2011a. "Macroeconomic and Operational Challenges in Countries in Fragile Situations." Washington, DC (June 15).

———. 2011b. "Managing Global Growth Risks and Commodity Price Shocks—Vulnerabilities and Policy Challenges for Low-Income Countries." Washington, DC (September 22).

———. 2011c. "Managing Volatility in Low-Income Countries—The Role and Potential for Contingent Financial Instruments." Washington, DC (October 31).

———. 2011d. "Revenue Mobilization in Developing Countries." Washington, DC (March).

———. 2011e. "Target What You Can Hit: Commodity Price Swings and Monetary Policy." In *World Economic Outlook,* ch. 3. Washington, DC (September).

Walsh, James P. 2011. "Reconsidering the Role of Food Prices in Inflation." IMF Working Paper 11/71, Washington, DC.

4

Using Trade Policy to Overcome Food Insecurity

Summary and main messages

Trade is an excellent buffer for domestic fluctuations in food supply. There is no global food shortage: the problem is regional or local—one of moving food, often across borders, from surplus production areas to deficit ones—coupled with affordability. World output of a given food commodity is far less variable than output in individual countries. Thus increased trade integration holds considerable potential to stabilize food prices, boost returns to farmers, and reduce the prices faced by consumers.

Trade liberalization protects national food markets against domestic shocks by allowing more food to be imported in times of shortage and exported in periods of plenty. However, historically—and despite a host of regional trade agreements—most countries have chosen to take the opposite approach by restricting imports of food and discouraging exports in often-failed attempts to keep domestic markets isolated from international shocks by ensuring self-sufficiency in food production.

Self-sufficiency should be weighed against the benefits of cheaper imports. A country that is a natural exporter should not hinder its comparative advantage with export bans. A country that tends to import food should allow its domestic market to remain linked to the world market. Food security therefore requires encouraging more trade through a more open, rules-based multilateral trade regime, best achieved by concluding the Doha Round of WTO negotiations, and supported by further work toward developing disciplines on export restrictions.

Efforts to extend trade integration to developing countries should also focus on promoting more effective regional integration among them, including for food products. Facilitating food trade is also important through increased Aid for Trade to promote frictionless borders and to induce a supply response from developing countries, particularly in Sub-Saharan Africa.

Trade in food

Global production of cereals has almost trebled in the past 50 years, outpacing the twofold rise in world population. Yet over a billion people in the world remain hungry. Cereals form the staple diet of poor people and are also their main imported food item. In 2010, cereals made up 40 percent of the food imports of least developed countries. Increasing consumption of vegetables and meat is indicative of growing incomes, and

FIGURE 4.1   Most cereal production is consumed domestically and not traded

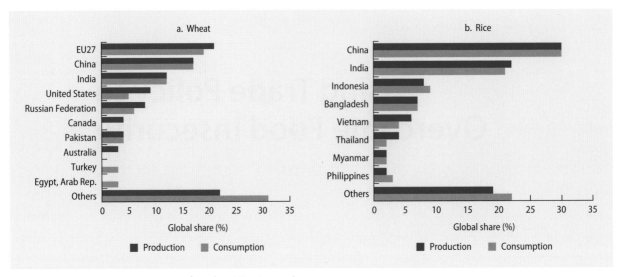

Source: Kshirsagar and Baffes 2011 (U.S. Department of Agriculture, 2006–10 averages).

these items typically account for half of the food imports by developed countries.

Wheat, maize, and rice account for the majority of trade in cereals; maize and other coarse grains are not only consumed by humans but are also used as animal feed in the production of meat and for the manufacture of biofuels. Most cereal production is for domestic consumption (figure 4.1), with just

10 percent of world production traded globally: over the past decade, only one-fifth of all wheat produced globally was traded, while rice trade accounted for 6 percent of global rice production (Kshirsagar and Baffes 2011).

In value terms, approximately two-thirds of world food exports go to developed countries, and just under one-third to middle-income ones, with the poorest countries being

FIGURE 4.2   Food trade matters most for low-income countries

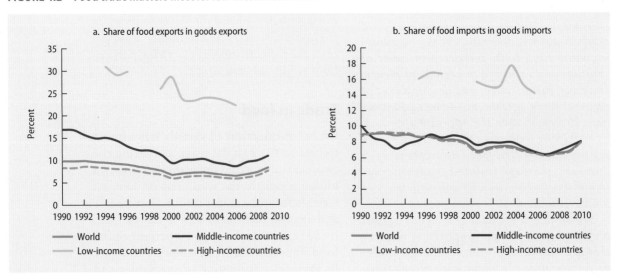

Source: World Bank DDP and COMTRADE data.
Note: Food = SITC rev. 4 codes 0+1+22+4: food, livestock, alcohol/tobacco, oilseeds, edible oil.

insignificant in world food trade: the share of the least developed countries in world food trade is just 1 percent. However, food trade forms a higher share of the *total* trade basket of developing countries compared with developed countries (figure 4.2) (FAOSTAT 2010 Yearbook). Sub-Saharan African countries, especially some in the Horn of Africa, also have high shares of food imports in total imports, compared with other parts of the world. Although not all developing countries depend on food imports, how food is moved within and across borders has clear implications for poor farmers and consumers, who spend a large share of their household income on food.

Markets in key cereals are often dominated by just a few players among developing countries (figure 4.3); India and China are the largest producers and consumers of these crops. Exports of wheat are mainly from developed countries, exports of rice from developing ones. More than 62 percent of all wheat is exported by the United States, the European Union (EU), Canada, and Australia, and these countries have highly protected agricultural sectors. South and East Asian economies are the leading rice exporters, but only 6–7 percent of global production is traded. Market concentration in cereals has declined over time, with an increasingly diversified export base, although the United

FIGURE 4.3   Trade in key cereals is dominated by just a few countries

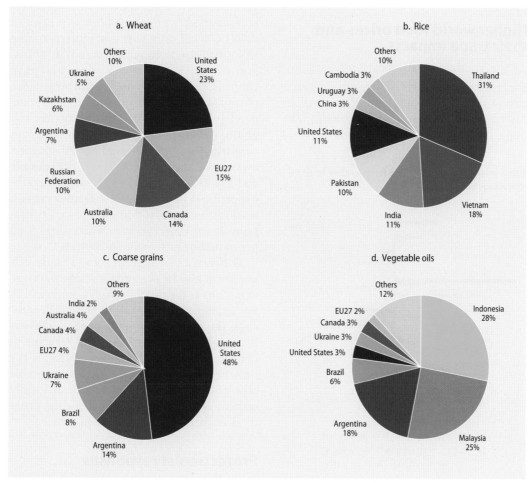

Source: Kshirsagar and Baffes 2011.
Note: U.S. Department of Agriculture, 2006–10 averages; coarse grains are those used as feed (maize, millet, sorghum, and barley).

States continues to dominate trade in maize (Kshirsagar and Baffes 2011). Import markets are, and have historically been, less concentrated than export ones.

Trade policy actions by exporting and importing countries can have knock-on effects in food markets, and food commodity prices are often highly correlated. For example, an export restriction by India on rice exports, even one that does not directly influence the world price, can still lead to market behavior that indirectly affects the world price, as happened in 2008 when other rice exporters also started to impose restrictions. Wheat, rice, and maize share a positive relationship: price changes due to temporary production or export disruptions can affect the price of substitute products (Ivanic, Martin, and Zaman 2011).

Higher world food prices and their trade impacts

Food prices remain at historically high levels, contributing to differing terms-of-trade effects across developing countries as well as distributional impacts within them. The impact of global food inflation on external

balances, growth, and welfare depend critically on the terms-of-trade effects of higher food prices. The increase in world food prices implies terms-of-trade gains for net-exporting countries of food products and losses for food-deficit, net-importing ones (figure 4.4). For example, net-food-importing countries in the Horn of Africa such as Ethiopia, Kenya, and Somalia currently face drought, famine, and humanitarian emergency situations affecting more than 13 million people, with domestic food prices soaring (between 30 and 240 percent for red sorghum and maize in Somalia), while Tanzania and Uganda have gained because they remain net exporters (mostly for maize).

Increases in global prices have not always translated into equivalent increases in food prices prevailing in domestic markets for various reasons, including a weakened dollar (commodity prices are often expressed in dollars); local transport costs (often arising from inadequate competition in road transport markets); market distortions and price controls set by governments; the persistence of trade barriers; and good harvests in some developing countries (notably for maize, sorghum, millet, and cassava in some African countries that have allowed for substitution away from imported wheat and rice) despite bad yields in several of the largest grain-exporting economies. These factors explain stark differences in domestic price fluctuations across countries even when world food prices decline or remain unchanged.

Differences in aggregate food trade balances can also be deceptive and conceal large variations at the product level (Canuto 2011). For example, in the Andean region, Venezuela is the only net importer of food whereas Bolivia, Colombia, Ecuador, and Peru are all net food exporters. However, Bolivia is the only net exporter of cereals and vegetable oils, whose price increases have dramatically spiked; coffee and bananas drive the other three countries' net exporting positions.

Protectionist responses

Protectionism should be avoided as global trade slows and food prices remain high.

FIGURE 4.4   Net-food-importing regions lose from higher food prices while net-exporting regions gain

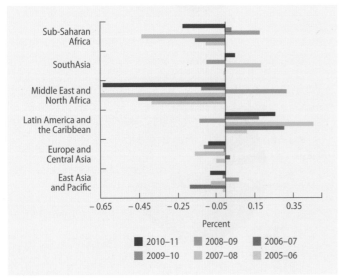

Source: World Bank, Datastream.
Note: Terms-of-trade changes in food commodities, by developing region, year over year changes as share of GDP.

Trade in food is currently subject to fewer policy interventions than has historically been the case, but since 2011 trade protection is once again increasing. Given renewed economic uncertainty in 2012, however, coupled with higher food prices and the tendency for countries to insulate their domestic markets from world price shocks, governments must continue to keep their markets open to avoid pushing domestic food prices higher.

For a number of staple food commodities, many governments intervene in their food markets in attempts to reduce the volatility of domestic prices relative to world prices. In developing countries, the various interventions reflect the sensitivity of governments to volatile prices for important staples, either to protect consumers against high prices or to maintain higher domestic prices for producers. Such measures can be shown to be second-best complements to storage policies for individual small and open developing countries concerned about the adverse impacts of high prices for staple foods on risk-averse consumers and farmers, when insurance against price volatility is unavailable and more direct measures to target poor households (in periods of high prices) and fragile producers (in periods of low prices) are not feasible (Gouel and Jean 2011). But such trade restrictions are not a cooperative way to address price volatility and can actually exacerbate it.

Trade restrictions have both direct and indirect impacts on world food prices. Trade-distorting policies displace and reduce the efficiency of agricultural production globally and make it less resilient to exogenous shocks: policies that distort production and trade in food commodities also potentially impede the achievement of long-run food security, by promoting production in areas where it would otherwise not occur and by obscuring the transmission of price signals to efficient producers elsewhere. Furthermore, a collective action problem may emerge: many countries simultaneously insulating their domestic markets against global price shocks through restrictive trade measures may well create higher volatility for global food prices (Martin and Anderson 2011).

Traditionally, it has been the trade policies of developed countries that were responsible for pushing down the world prices of agricultural products, including those exported by developing countries. However, over the past two decades there has been a shift in agricultural protection to developing countries, with reductions in export taxes but increases in protection on import-competing goods. Tariffs on food trade are highest for goods from middle- and high-income countries, averaging 22 percent (Boumellassa, Laborde, and Mitaritonna 2009). In developed countries, agricultural protection remains high but has declined from its peak level during the 1980s. While lowering global protection can be expected to raise demand and therefore increase world food prices by a relatively small degree, global trade liberalization is actually likely to lower prices faced by consumers in developing countries, with the rise in world prices offset by reductions in domestic ones.

Cooperative options to lowering domestic food prices therefore include *permanently* reducing import tariffs and other taxes on key staples and agricultural inputs. Instead, however, countries often tactically lower import barriers on food *temporarily* during periods of domestic food scarcity only to reimpose them later when yields have improved, again exacerbating world price volatility (Martin and Anderson, 2011). "Water" in the tariff (the difference between bound and applied rates) can leave significant room for countries to raise their applied tariffs on food imports, also compounding global price volatility. Lowering bound tariffs has been a core part of the Doha agenda.

Other trade measures such as export restrictions and non-tariff measures (NTMs), including domestic policies such as price support, also influence the extent to which price changes in domestic markets accurately reflect world prices. The World Trade Organization (WTO) reports that trade restrictions over the past year have spiked, particularly since July-August 2011 when the debt crises in the Euro Area and the United States began to intensify (WTO 2011b). Protection measures by the Group of 20 (G-20) countries—the main users of trade restrictions—now affect a little over 2 percent of world trade. Approximately 1,000 trade-restrictive measures were introduced between September 2008 and October 2011, with increasing use of NTMs, especially quantitative import restrictions (Datt, Hoekman, and Malouche 2011). One-third of all NTMs were on exports, with increased use of export restrictions for agricultural products, in part as a result of higher world food prices.

Since September 2008, new trade-restrictive measures on food products (that is, all products within SITC Rev. 4—food and live animals, beverages and tobacco, oilseeds and edible oils), has accounted for one-quarter of all new trade restrictions, and the share is rising. Export restrictions have been used in attempts to stabilize domestic food prices (figure 4.5). But these same policies have exacerbated global food price volatility, raising the price of rice by 45 percent and that of wheat by almost 30 percent between 2006 and 2008 (Martin and Anderson 2011). New trade restrictions adopted between September 2008 and October 2011 were applied

most frequently to meat, livestock, and grains (concern over pandemics drove the restrictions applied to livestock). The most frequent users of protection measures for food over the period were China, India, Indonesia, and the Russian Federation, which together accounted for almost one-third of all trade restrictions introduced on food items since the beginning of the financial crisis. Non-G-20 countries, most notably Belarus, Bolivia, and Ukraine, have also imposed trade restrictions on food products.

Notably, since the 2008 financial crisis, countries have also pursued trade liberalization as well as protection in efforts to lower domestic prices for households and industries (figure 4.6). Although some countries have increased their import tariffs on food products—for example, Russia increased its tariffs to 50–80 percent on imports of pigs, pork, and poultry—tariff reductions on food imports were far more frequent over this period. In some cases the reductions in import tariffs were significant. For example, Turkey lowered its tariffs on livestock from

FIGURE 4.5   **The most frequent users of trade-restrictive measures on food products are G-20 countries**

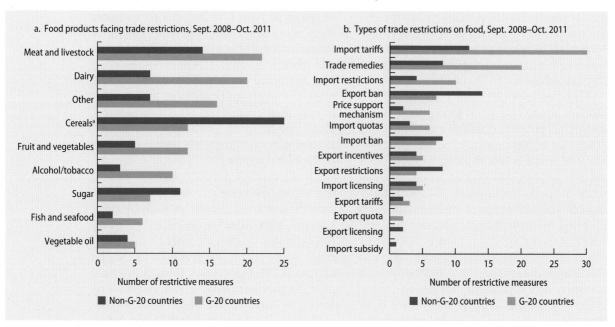

Source: Authors' calculations using data from WTO Trade Monitoring Reports, 2009, 2010, 2011.
Note: Total restrictions = 177; restrictions depicted exclude pandemic-related measures; trade remedies = antidumping, countervailing duties, safeguards.
a. In G-20 countries, "cereals" are mainly wheat; in non-G-20 countries, "cereals" are mainly wheat and rice.

FIGURE 4.6   Some countries have also sought to lower domestic food prices by temporarily lowering trade restrictions

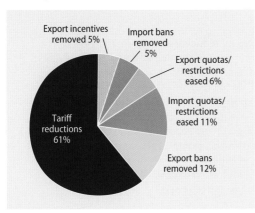

Source: Authors' calculations using WTO data from 2009, 2010, 2011.
Note: Trade liberalizing measures on food products, September 2008–October 2011; total number of observations = 148.

FIGURE 4.7   Producer support to farmers in most developed countries has fallen but is rising in emerging economies

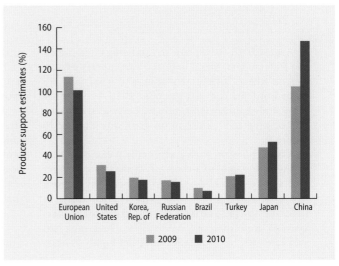

Source: OECD 2011.

135–225 percent to 0–20 percent. Most food tariff reductions were on grains and sugar, followed by meat, edible oil, and dairy products. Additionally, some countries have tried to stimulate exports with various incentives: Brazil, through duty drawback schemes on meat exports; and the European Union and the United States with refunds and other incentives to their dairy industries.

Direct subsidies to farmers in developed countries remain a major source of support, disadvantaging producers in other countries and distorting world trade. Producer support estimates (PSEs) produced by the Organisation for Economic Cooperation and Development (OECD) provide a measure of the extent to which developed country governments are assisting their farmers over time through various payments and price support policies. PSE expresses the monetary value of policy transfers from consumers and taxpayers to producers and can also be expressed as a percentage (%PSE) of gross farm receipts. Support to producers in developed countries was estimated to be $227 billion in 2010, accounting for 18 percent of gross farm receipts—the lowest %PSE on record (OECD 2011).

PSEs have increased in China, Japan, and Turkey (figure 4.7). In China support has been increasing rapidly and is actually nearing the OECD average %PSE. For OECD countries, rice, sugar, milk, and livestock receive the highest level of support through price protection policies and payments based on output, although large declines in price support in recent years have been associated with high world prices for these products. Milk, sugar, and rice also feature prominently among the commodities receiving specific support in emerging economies. As mentioned in chapter 1, biofuel policies in developed countries, which consist of subsidies, tax credits, and legislative mandates, have further distorted agricultural trade.

Developing countries also use policies that adversely affect food trade and are highly restrictive; such measures include food marketing boards, oligopolistic market structures in key parts of the food value chain such as milling, price controls, and trade bans. Countries that are net exporters of food may face political pressures to restrict food exports in periods of high domestic prices. Not only do these policies tend to have a limited impact on domestic price levels, however, but they also can have a significant negative effect on earnings from export production (box 4.1). Countries that insulate their domestic markets also export instability onto international markets,

BOX 4.1   Russia's export ban on grains

In August 2010, in response to escalating grain prices, the Russian Federation imposed a temporary export ban on wheat, barley, rye, maize, and wheat and rye flour until the end of December 2010. In October 2010, the export ban on grain was extended until the end of June 2011; the ban on flour was allowed to expire.

The export bans were originally a response to a drought that caused a shortfall in the grain harvest and associated rapid grain price increases in both domestic and international markets. According to official estimates, farmers harvested almost 37 percent less grain than they did in 2009. The export ban was intended to insulate Russia from highly volatile grain prices by reducing exports in 2010–11 to the 3 million tons already shipped at that time, resulting in a drop of nearly 12 million tons of exports initially projected for the year.

The export restrictions had unintended and undesirable consequences such as undermining Russia's long-term policy of becoming an even more important player in the global grain market, encouraging hoarding in expectation of the bans' removal, distorting prices, and affecting the investment and production decisions of its farmers.

Source: World Bank 2011d.

especially if they are major producers or consumers of food. For example, the introduction of export restrictions on food exports by Argentina, Kazakhstan, Russia, and Ukraine for wheat and China and India for rice, in attempts to decouple domestic markets from global markets to keep domestic food prices low, have in the past compounded the food price problem.

Smaller developing countries (such as Malawi, Tanzania, and Zambia) also routinely impose strict controls on food trade, especially if their agricultural sectors remain highly regulated by various interventions at local and national levels. For example, some countries often ban imports during good harvest years to ensure domestic production is consumed first and limit exports during periods of low yields to contain domestic price increases. While these policies are often implemented ostensibly to promote food security in the form of self-sufficiency, they rarely work and can exacerbate food insecurity rather than reduce it (box 4.2).

Some restrictive barriers to trade are not always as visible as outright bans but come in more nebulous, less apparent forms that nevertheless increase trade costs.

Trade costs between Maghreb countries in North Africa—Algeria, Libya, Mauritania, Morocco, and Tunisia—are two to three times higher than those faced between countries just north of the Mediterranean rim (such as France, Italy, and Spain). This differential is partly attributable to more NTMs and constraints to intraregional trade versus interregional trade, such as more border controls and limited cross-border cooperation to facilitate trade across land borders (box 4.3). These regional barriers to trade drive up the costs of trading agricultural products, with significant implications not only for food security, but for political stability and economic development more generally.

The persistence of NTMs on trade in food reduces trade in these products. New research at the World Bank suggests that the ad valorem equivalent of NTMs on African cross-border trade in food is very high (Gourdon and Cadot 2011). For example, sanitary and phytosanitary (SPS) regulations on imports of rice raise prices by as much as 42 percent in Kenya and 30 percent in Uganda (box 4.4).

Bans and other restrictions on food trade as well as government interventions that

BOX 4.2   Government imports of maize during the Southern Africa food crisis

In 2001–02, the Zambian government publicly announced that it would import 200,000 tons of maize from selected South African suppliers to cover the national food deficit and sell it below market price to a small number of large formal-sector millers. The subsidy was intended to limit consumer price increases, paid directly to the South African suppliers and also to the importers. Because of liquidity problems, the subsidy payment was made late, causing the maize imports to be delayed. When the government instead imported only 130,000 tons very late in the season, maize and maize flour shortages occurred and local market prices exceeded import parity. Zambian traders and millers who had not been selected to benefit from the scheme, including informal traders from Mozambique, refrained from commercially importing maize for fear of not being able to sell it once the subsidized maize reached the market. Because grain was channeled only to the largest millers, consumers had to pay a higher price for already-refined flour instead of being able to source grain and mill it themselves or though the informal network of small hammer mills.

In the same year, Malawi also faced a modest maize production deficit—8 percent below the country's 10-year average. In September 2001, its grain-

trading parastatal (ADMARC) announced a fixed price for maize sold at its distribution centers and declared its intention to import maize from South Africa to maintain this price. The selling price was set considerably lower than the landed cost of imported maize, leaving private traders with no incentive to commercially import. As with Zambia, the government imports also arrived late and were insufficient to meet demand, so prices soared to a peak of $450 a ton in early 2002. To make matters worse, the late-to-arrive ADMARC imports arrived during the good 2002 harvest and were then released to the market, resulting in 16 months of continuously falling maize prices, to the detriment of farmers. At other times, the sourcing of grain from South Africa and subsequent release onto the domestic market through government contracts with South African suppliers has also depressed informal maize trade with Mozambique. Because Mozambique is the source of informal trade in maize to southern Malawi, these government imports also add greater risks and price instability for Mozambique's smallholder farmers.

Sources: Nijhoff et al. 2002; Jayne, Chapoto, and Govereh (2007); Rubey 2005; Nijhoff et al. 2003.

foment distortions might allow a country to shield consumers from the initial implications of a price hike. However, they do not provide the incentive for a domestic supply response, and these implications should be considered when implementing policies that restrict international trade. Encouraging more trade in food, not less, is essential for achieving food security. Increased reliance on trade for production and consumption of food, as well as for inputs, increases farm gate prices without necessarily inflating consumer prices—a win-win for farmers and consumers alike. Indeed, those developing countries that have adopted more-open trade policies for food have seen benefits through higher production, exports, and trade in these products, together with lower domestic price volatility (box 4.5).

However, national self-sufficiency in food remains a highly sensitive issue in both developed and developing countries in which political struggles are sometimes played out in food marketing and trade policies. Price shocks on net-food-importing countries can also widen current account deficits, put additional pressure on exchange rates, cause a shortage of foreign reserves, and increase social safety net expenditures. For example, during the Arab Spring, the government of Jordan overturned the food subsidy cuts it made in 2008 and introduced tax exemptions on 13 foodstuffs. In the Arab Republic of Egypt, the bread subsidy is estimated to reach around 85 percent of the population (World Bank 2011a). The risk with such measures, however, is that they can become entrenched, incurring high fiscal costs.

BOX 4.3   The Middle East and North Africa region faces high trade costs in food

The Mediterranean basin, including its northern European and southern North African rims, has been an active trading area for more than three millennia. Yet trade and logistics patterns between the two rims vary considerably, with the cost of trading among Middle East and North African countries being inordinately high. Trade costs between countries on the developing, southern rim are higher than those experienced between the wealthier, European counterparts (such as France, Italy, and Spain), by as much as three times for agricultural goods. Moreover, trade costs *within*, for example, the Maghreb region or between the Levant countries in the Eastern Mediterranean exceed those the region incurs *externally* with Europe. Three explanatory factors stand out, in order of restrictiveness: NTMs that constrain trade processes; the low quality and fragmentation, by country, of logistics services such as trucking; and less developed intraregional infrastructure, such as ports that easily connect the Maghreb to the Mashreq, and few active transport corridors between countries. (Trucking and railway movement are still suspended or heavily controlled at several borders in part because of security concerns but also because of mutual lack of trust regarding standards or origin, especially in the context of the Pan Arab Free Trade Agreement, which will remove tariffs on all goods of Arab origin.) A 2009 World Bank mission counted as many as 10 separate control stops at the Syria-Jordan border, equally distributed on either side. Container dwell time in Morocco and Tunisia is about one week, longer than the OECD benchmark of 3 days and that in emerging Asia— 4 days in Malaysia, 2.5 days in Shanghai. Small reductions in trade costs can result in considerable trade expansion: reducing trade costs by just 5 percent could increase trade between the Maghreb and Western Europe by 22 percent, and intra-Maghreb trade by 20 percent. Lower trade costs would also facilitate production sharing within a larger market resulting in more competitive exports to Europe.

Sources: Shepherd 2011; Arvis 2012; Hoekman and Zarrouk 2009.

BOX 4.4   Quantifying the effects of non-tariff measures on trade in African food staples

Quantifying the price-raising effect of non-tariff measures (NTMs) was, until recently, constrained by the availability of comparable data across countries. Thanks to a collaborative effort between the World Bank and other agencies, including United Nations Conference on Trade and Development (UNCTAD) and the African Development Bank, a new wave of data collection was undertaken in 2009–10. So far, 30 countries have been covered, with NTMs coded for each of the Harmonized System's 5,000 product lines. Combining this data with price data collected as part of the World Bank's International Comparison project (for a smaller set of products) has made it possible to estimate directly, using econometric methods, the price-raising effect of NTMs on African food staples.

The approach consisted of running regressions of country-level product prices on "dummy" (binary) variables marking the application of NTMs of various types, using a panel of 1,260 country-product pairs. The regressions control for systematic differences in cost-of-living across countries, as well as in market-structure diversity across products, with a full array of country and product fixed effects. Interaction terms between NTMs and either region or country dummies provide tentative estimates of their price-raising effect in Africa or in specific countries.

As is usual with this type of exercise, results should be interpreted with caution, because many confounding influences can affect estimates. Although many controls are used in the regressions to limit these confounding influences, they put heavy demands on the data and result in many coefficients being estimated with large confidence intervals. Be that as it may, results, shown graphically in the figure below, are telling. On average Africa's SPS measures, which often

BOX 4.4   Quantifying the effects of non-tariff measures on trade in African food staples (continued)

Price-raising effect of NTMs, Africa average (all affected products)

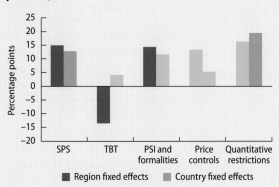

Note: SPS = sanitary and phytosanitary; TBT = technical barriers to trade; PSI = pre-shipment instructions.

suffer from lack of harmonization, poor design, and haphazard enforcement, raise the price of food staples by 13–15 percent. Quantitative restrictions, where they are applied, add another 20 percent. Such price increases have the potential to affect significantly the real income of poor households.

Product-specific estimates suggest substantial effects of SPS regulations in Kenya on rice prices (+42 percent), meat (+34–37 percent), fish (+33 percent), and edible oils and fats (+29 percent). Rice prices seem to be similarly affected in Uganda (+30 percent), as are meat and fish prices (+41 percent).

Source: Gourdon and Cadot 2011.

BOX 4.5   Open border policies for trade in food

Unlike many other countries in the region, Uganda and Mozambique have consistently retained liberal border policies for food staples. Uganda's open trade policy for food staples enables traders to offer products and services competitively, reliably, and on a sustainable basis. Uganda is able to serve as a food basket for East Africa. There is no export restriction on agricultural products, nor has the government instituted any recent ban on trade in food. Consequently, the flow of maize from Uganda to Kenya is one of the larger and more consistent cross-border flows in the region (approximately 120,000 tons a year). There is also cross-border trade with Rwanda (50,000 tons), and southern Sudan is also becoming a growth market for Ugandan products. Nevertheless, the most distinct feature of the Ugandan market is the significant presence of the World Food Programme (WFP) and its procurement program. Ugandan maize accounts for the largest proportion of maize the WFP procures in Africa (21 percent in 2010), excluding South Africa (which accounted for 24 percent in 2010). The WFP buys Ugandan maize as well as beans for distribution to internally displaced people in the country but also sends shipments to Burundi, the Democratic Republic of Congo, Kenya, Rwanda, Sudan, and Tanzania, all

of which periodically face food shortages. The volumes purchased reached 109,000 tons in 2010. The maize policy of the Ugandan government has been to allow and encourage cross-border trade, and the WFP procurement program has encouraged a supply response of more maize and beans from farmers who are able to meet WFP's quality and quantity requirements.

Mozambique, since the end of its civil war in 1992, has also freely allowed both imports and exports of maize. Because northern Mozambique is typically a maize surplus area, and because Malawi offers better prices than southern Mozambique (because of longer distances and higher transport costs to Maputo), traders in northern Mozambique routinely sell their grain to Malawi and eastern Zambia. The open border policy enables the resulting deficits in Mozambique's southern cities to be met by large millers who import grain from South Africa and mill it for domestic sale. Trade (coupled with the 30 percent subsidy on flour for wheat and bread production) has therefore helped to stabilize prices in Maputo compared with other capital cities in the region.

Sources: Haggblade et al. 2008; World Bank 2009a.

Moreover, consumer subsidies that are met by price controls and trade restrictions can be counterproductive and create disincentives for domestic food producers.

Policy recommendations for opening food trade in the pursuit of food security

Concluding the Doha Round would bring more predictable market access for food products

A conclusion to the Doha Round of WTO negotiations would contribute to food price stability by reducing distortions and strengthening disciplines on food trade restrictions, thereby limiting the scope for countries to impose policies that destabilize world food markets. It would also provide a boost to the world economy, generating a potential stimulus of $160 billion in real income (Laborde, Martin, and van der Mensbrugghe 2011). The primary deliverable would be enforceable policy commitments by member governments to provide greater security of market access by not raising support for domestic agricultural sectors above a given level (high commodity prices could dissipate resistance by farmers in developed countries to an agreement on this); to place greater restrictions on the level of permitted tariffs for food imports; and to refrain from using certain policies at all (such as export subsidies). The topics on the table are therefore important, and in principle there is enough substance for all countries to gain from an agreement. However, too much emphasis has been placed on the gains from market access alone. The Doha Round is about much more than market access. Concluding the negotiations arguably requires greater recognition of the value that new trade policy disciplines could bring as part of an agreement (Hoekman 2011). For example, while a complete ban on export subsidies for crops such as cotton would be a major step forward, it should not be quantified by estimating the impact

of removing extant subsidies—especially in a period where high prices have reduced the prevalence of their use. The ban would be more significant if world prices fell in the future because the decline would not trigger an increase in export subsidies.

Developing disciplines on the ability of governments to use import and export barriers to insulate domestic markets, and hence make world markets less thin, would also be a major source of welfare gain for developing countries (Martin and Anderson 2011). WTO disciplines for food export restrictions are currently considered to be insufficient and weak (FAO and OECD 2011). Export taxes are covered under the WTO and must comply with the most-favoured-nation clause. Article XVIII of the General Agreement on Tariffs and Trade (GATT) also provides for the negotiation of tariffs on both imports and exports. And while export taxes do not have an institutional anchor equivalent to import tariff bindings, which are addressed specifically in the GATT (Article II), there is no legal impediment to negotiating their reduction or elimination (Mavroidis 2007). However, there are very few export tariffs that have already been negotiated by WTO members. This means that most export tariffs are not yet bound.

Quantitative restrictions, including for exports (bans), are generally prohibited by Article XI of the GATT but an exception allows members to restrict food exports as long as the measures are "temporarily applied to prevent or relieve critical shortages of foodstuffs. . . ." (GATT Article XI, 2 (a)). Export restrictions relating to food must also conform with the Agreement on Agriculture (Article 12) that requires WTO members to maintain transparency in using such measures by considering the effects on importing members' food security, give notice in writing, and consult with other WTO members as far in advance of implementation as is possible. The provisions of this article exempt developing countries, unless they are net food exporters of the specific food staple concerned. However, since June 2010, only four notifications by three WTO members have been submitted (Saez 2011).

One policy option, therefore, would be to ban export restrictions altogether in the WTO if this could be agreed and enforced. Commitments by the larger exporters of food not to impose export restrictions would especially help maintain world price stability in periods of food stress. Reinforced multilateral trade rules for notification and transparency of export restrictions would also be useful. Developing a code of conduct to exempt food aid from export restrictions is an important priority for the international community. Food and Agriculture Organization (FAO) member countries have already agreed to remove these on food consignments purchased for humanitarian purposes, first at the Group of Eight (G-8) Summit in L'Aquila, Italy, in July 2009 and then at the World Summit on Food Security in Rome in November 2009. If met, these commitments would allow food to be shipped to where it is most needed in times of severe shortage.

Greater opening of regional markets to trade would promote food security and price stabilization

The potential for faster agricultural growth in many developing countries could be unlocked by deeper regional trade integration to complement multilateral liberalization efforts. In the absence of a Doha package, increased regional trade can also be a powerful instrument for stabilizing food supply and food prices. The distribution of food crop cultivation between neighboring countries, coupled with possibilities, where they exist, for staggered harvesting within the same commodity, offers substantial opportunities for regional trade. Because production variability is not often highly correlated among countries in most regions, integration through regional trade can reduce the effects of small country size on production volatility.

Examples of regional trade in food, both recorded and unrecorded, are numerous and include northern Zambia, where cassava

production ensures domestic food security, even in drought years, enabling the region to export maize to the Democratic Republic of Congo, Malawi, and elsewhere in Zambia; eastern Uganda, where bananas and cassava ensure food security, thereby enhancing maize exports to chronically food-deficit Kenya; northern Mozambique where cassava and Irish potato cultivation provide local food, enabling regular maize exports both north into Kenya and south into Malawi; most of Tanzania where a combination of rice, cassava, bananas, and maize enable regular cereal exports both north into Kenya and south into Malawi; and South Africa where large-scale commercialization and mechanization combined with modern inputs and irrigation enable high yields for the export of cereals northward to Zimbabwe, southern Mozambique and Malawi (Haggblade 2008). Indeed the scope for increased cross-border trade in Africa is enormous, but various obstacles remain (box 4.6). Elsewhere, Thailand, the world's largest producer of cassava, has recently witnessed dramatic increases in its exports of this crop on the back of sales to China for biofuel production.

To better exploit these opportunities, more effective regional trade policy and regulations must be developed to link smallholder farmers to urban demand centers across borders. Groups of developing countries have been actively pursuing regional trade agreements (RTAs), including the formation of free trade areas and customs unions, which for the most part have largely succeeded in reducing *tariffs* on most goods traded among them. As with global trade, however, the gradual removal of tariffs has meant NTMs have become more visible. For example, export bans, country-specific standards, complex rules of origin, and cumbersome customs requirements across countries often serve to reduce regional trade and destabilize regional food prices.

Additionally, governments have retained the use of safeguards under their various RTAs to exclude food from open regional trade on the grounds of health and public safety. As a result, governments retain a

---

BOX 4.6   Defragmenting Africa: What will stimulate regional trade integration?

Africa's potential for regional trade remains unexploited because of the high transaction costs that face those who trade across borders in Africa. A wide range of policy-related barriers drives up costs and limit trade. To escape the current straitjacket of trade fragmentation, Africa needs to pursue changes in three key areas:

- Facilitating cross-border trade, especially by small poor traders, many of whom are women, by simplifying border procedures, limiting the number of agencies at the border and increasing the professionalism of officials, supporting traders' associations, improving the flow of information on market opportunities, and assisting in the spread of new technologies, such as cross-border mobile banking, that improve access to finance.
- Removing a range of nontariff barriers to trade, such as restrictive rules of origin, import and export bans, and onerous and costly import and export licensing procedures.
- Reforming regulations and immigration procedures that limit the substantial potential for cross-border trade and investment in services.

The main message is that to deliver integrated regional markets that will attract investment in agroprocessing, manufacturing, and new services activities, policy makers need to move beyond signing agreements that reduce tariffs to drive a more holistic process to deeper regional integration. An approach is needed that reforms policies that create nontariff barriers; puts in place appropriate regulations that allow cross-border movement of services suppliers; delivers competitive regionally integrated services markets; and builds the institutions that are necessary to allow small producers and traders to access open regional markets. The appropriate metric for successful integration is not the extent of tariff preferences but rather reductions in the level of transaction costs that limit the capacity of Africans to move, invest in, and trade goods and services across their borders.

While there have been many initiatives to integrate regional markets in Africa, effective implementation of commitments has been sorely lacking. Hence, there is a need to help countries understand the political economy behind resistance to integrative reforms. How is it that leaders publicly and, by and large, genuinely pledge support for integration, but actual barriers to trade persist? For example, most of the nontariff barriers identified in the East African Community for immediate removal in 2008 are still in place. Opening up food staples to regional trade will create winners and losers. Therefore, political consensus on agricultural reform is required to create new institutional arrangements that moderate the impact of future shocks and instability in agricultural markets. Two related factors can help governments build constituencies for reform and provide a predictable and stable policy environment:

1. *An inclusive dialogue on food trade reform informed by timely and accurate data on global, regional, and national markets.* Food trade policy is rarely subject to open discussion, and the interests and views of relevant stakeholders in food staples trade policies are seldom represented. And when there is open discussion about trade reform, decision makers rely most on the input of those with political influence.
2. *A reform strategy that provides a clear transitional path to integrated regional markets rather than a single but politically unfeasible jump to competitive markets.* A reform strategy will have to take place in incremental steps that encourage investment by reducing uncertainties about policies for the private sector and deliver real and visible benefits. At the same time, it will allow policy makers to move at a pace consistent with their political risk calculations and their capacity to address the concerns of those who will lose from the reform process.

Source: World Bank 2012.

---

great deal of discretion over food-related trade policy, particularly in cases of food security and when there is a perceived risk to human health. Consequently, regional trade policy for agricultural products has essentially become a patchwork of rules implemented unevenly across different countries and enforced inconsistently, generating an opaque policy environment that severely limits trade in food.

Improved transport logistics and trade facilitation would improve links to markets and promote cost-effective access to food and food inputs

Trade policy restrictions are not the only impediment to the free movement of food across borders. Efficient transport and logistics are critically important to agricultural marketing and are a key component of prices. Yet in developing countries, particularly landlocked least developed countries, transport and logistics costs are generally far higher than OECD benchmarks of around 9 percent. For example, on average transport and logistics costs account for 18 percent of the value of firms' sales in Latin America, reaching 32 percent for Mercosur (Southern Cone Common Market) and Chile (World Bank 2005). In the case of African countries, improvements in logistics services (as measured by the Logistics Performance Index) would provide greater benefits than changes in other components of trade costs (Hoekman and Nicita 2008).

Transport and logistics costs are also an important determinant of food costs for importing countries as well as of food price variations within them. For example, maize prices in Guatemala have increased significantly more than in the rest of Latin America because of higher transport costs. Similarly, sharp increases in the prices of wheat-related products in Azerbaijan, the Kyrgyz Republic, and Tajikistan over the past year partly reflect increased transport costs from Kazakhstan (World Bank 2011b). While individual countries cannot do much to reduce ocean freight costs, which may be a significant part of the final price for bulk, relatively low-value commodities such as grains and edible oils, they can pursue proactive policy initiatives to lower costs associated with regional and domestic distribution. Investments in transport infrastructure have a proven track record of reducing consumer prices, especially in remote locations such

as Nepal. However a stronger focus on the "software" (regulatory) dimensions of transport, logistics, and trade facilitation projects is also needed (Arvis, Raballand, and Marteau 2010).

Improving trade facilitation and logistics reforms, as well as streamlining regulatory frameworks in the context of simplified border management procedures, can have significant benefits for consumers, while generating a favorable supply response. When moving formal consignments of food across borders, traders in developing countries often face a host of repetitive fees, permissions, redundant documentation procedures, and uneven certificate of origin requirements. As a result, customs clearance in many developing countries involves long delays, even for perishable goods such as food that should be cleared quickly. Individually most of these requirements may constitute a small delay or expense to traders but collectively they represent a significant barrier to trade. Even where single entry documents have been introduced, the information and accompanying documents (such as import declaration forms, origin certificates, invoices, import permits and standards compliance) required from traders can be burdensome, and small cross-border traders may be unable to provide all of the information for the entry document. For example, in Tanzania all certificates and permits can be obtained only in person in Dar es Salaam. In Kenya permits to legally import grain are available only in Nairobi (Nyameino, Kagira, and Njukia 2003). And traders wanting to export food staples from northern Mozambique to southern Malawi are required to get an export permit from Quelimane on the central coast of Mozambique (Tschirley, Abdula, and Weber 2005). Consequently, food trade can be effectively prohibited; subjected to tariffs (even if undertaken within the context of an RTA); or, as already discussed, pushed into informal channels.

Simple, structured, stable, and predictable trade regimes are needed that are based around harmonized and easy-to-satisfy border procedures that reflect the capacities of farmers and traders; the provision of information on rules and regulations that are easily available and well known; and clear notification procedures for new rules and regulations that allow traders, other governments, and agencies to contest proposed changes and give producers time to adjust.

Increasing the productivity of food production also requires an assessment of the problems that affect the whole value chain, particularly those relating to infrastructure and links to markets. The prices that farmers receive and consumers pay for food are influenced by the quality and availability of a range of services including extension services, transport and logistics services, storage and distribution, and water. Increasing competition in these services can play a positive role in boosting agricultural productivity and improving cost efficient access to food.

Positive policy measures to promote food security should be developed though increased Aid for Trade

Policy makers are often reluctant to open up to food trade because they are keenly aware that food price shocks can lead to food insecurity and consequently to social unrest. This is certainly the case if at the country level no social safety nets or other instruments are available to mitigate the adverse effects on the poor and vulnerable. At the same time, it is not always immediately clear whether a food price shock is permanent or transitory. Policy makers often treat shocks as transitory and use trade policies to protect their consumers. Those policies do not necessarily provide incentives to producers to increase productivity and production. As various improvements in the food value chain will require time to materialize, for example, in trade-related infrastructure, it is important to work simultaneously on enhancing social safety nets.

While rising world food prices are currently perceived as a "crisis" and are clearly a burden to poor net consumers of food, over the long term, they could bring significant opportunities to stimulate food production in developing countries, thus improving food security for the poor. They could also

enhance the contribution of agriculture to economic growth through attracting investments in agricultural research and more productive agricultural techniques, thereby harnessing the gains for small-scale farmers as well. Countries such as Brazil, Malaysia, and Thailand have made significant progress in agricultural commercialization in recent years and have undertaken investments in research and extension services while other countries such as India and Mali have improved their market information systems (World Bank 2009b). However, exploiting these opportunities requires an open and predictable trade policy environment for food and food inputs. For example, those policies that seek to control domestic food markets through price controls, direct government involvement in marketing activities, and trade restrictions are all likely to lower the food supply response over the medium term. In contrast, the development of market-based mechanisms to manage food price risks (such as futures and options markets, facilitation of private storage or warehouse receipts systems, market information systems, and weather-indexed insurance) are all likely to mobilize significant new investments from the private sector. Aid for Trade could be used to support the policy reform and supply-side upgrade processes that are needed for developing countries to better tap the opportunities created by more open multilateral and regional markets for food.

In 2009 Aid for Trade commitments reached approximated $40 billion—a 60 percent increase from the 2002–05 period. The share of Aid for Trade going to least developed countries has also increased from 26.5 percent during the period 2002–05 to 30.4 percent in 2009. Furthermore, support for multicountry programs (both global and regional) reached $7 billion in 2009, more than three times the amount during 2002–05. The World Bank is the largest provider of Aid for Trade. Based on the OECD/WTO definition of Aid for Trade, the Bank provided an average of $15 billion a year in Aid for Trade between 2001 and 2011 and accounts of 20 percent of all Aid for Trade expenditures globally. Lending for transport infrastructure is a critical component of the World Bank's efforts to help developing countries achieve their trade integration and policy reform objectives. Almost two-thirds of World Bank support for transport infrastructure is for roads and highways, with South Asia being the largest recipient of funds for transport projects. Excluding infrastructure, the World Bank provided a total of $2.6 billion in trade-related lending in 2010–11, an almost fivefold increase over 2002–03; the share of trade-related lending in total Bank lending also showed a rising trend, from an average of 2 percent during 2001–03 to an average of 6 percent during 2008–11. Africa is the largest recipient of World Bank Aid for Trade and now accounts for more than one-third of disbursements.

With uncertainty in the global economy and fiscal pressures in key donor countries, a key challenge will be to sustain current levels of financing. Monitoring by the OECD and WTO as part of the self-assessment exercise for the Third Global Review of Aid for Trade indicates that the outlook for Aid for Trade appears stable, although the previously high growth rates have declined. Aid for Trade grew by 2 percent between 2008 and 2009, compared with annual increases of 10 percent between 2006 and 2008 (WTO 2011a). Existing Aid for Trade pledges should therefore be honored and new pledges encouraged.

References

Arvis, J. F. 2012. "Trade and Transport Facilitation in the MNA Region." World Bank, Washington, DC.

Arvis, J. F., G. Raballand, and J.F. Marteau. 2010. *The Cost of Being Landlocked*. Washington, DC.

Boumellassa, H., D. Laborde, and C. Mitaritonna. 2009. "A Picture of Tariff Protection across the World, MAcMap-HS6, Version 2." IFPRI Discussion Paper 903, International Food Policy Research Institute, Washington, DC.

Canuto, O. 2011. "Fiscal Consequences of Food and Agricultural Commodities Inflation." Remarks for the World Bank GAIM/GMA,

Geopolitical Risk, Macroeconomics and Alternative Investment Conference, New York, October 11–12.

Datt, M., B. Hoekman, and M. Malouche. 2011. "Taking Stock of Trade Protectionism since 2008." Economic Premise Note 72, World Bank, Washington DC.

FAO (Food and Agriculture Organization). 2009. "The State of Food Insecurity in the World." Rome.

———. 2010. FAO Statistical Yearbook 2010.

FAO and OECD (Organisation for Economic Co-operation and Development). 2011. "Price Volatility in Food and Agricultural Markets: Policy Responses." Report prepared for the G-20. Rome and Paris.

Gouel, C., and S. Jean. 2011. "Optimal Food Price Stabilization in a Small Open Developing Country." CEPII (Institute for Research on the International Economy), Paris.

Gourdon, J., and O. Cadot. 2011. "The Price-Raising Effects of Non-Tariff Measures in Africa." World Bank, Washington, DC.

Haggblade, S. 2008. "Enhancing African Food Security through Improved Regional Marketing Systems for Food Staples." Michigan State University.

Haggblade, S., J. Govereh, H. Nielson, D. Tschirley, and P. Dorosh. 2008. "Regional Trade in Food Staples: Prospects for Stimulating Agricultural Growth and Moderating Short-Term Food Security Crises in Eastern and Southern Africa." Paper prepared for World Bank, Washington, DC.

Hoekman, B. 2011. "The WTO and the Doha Round: Walking on Two Legs." Economic Premise 68, Poverty Reduction and Economic Management Network, World Bank, Washington, DC.

Hoekman, B., and A. Nicita. 2008. "Trade Policy, Trade Costs and Developing Country Trade." Policy Research Working Paper 4797, World Bank, Washington, DC.

Hoekman, B., and J. Zarrouk. 2009. "Changes in Cross-Border Trade Costs in the Pan-Arab Free Trade Area, 2001–2008." Policy Research Working Paper 5031, World Bank, Washington, DC.

Ivanic, M., W. Martin, and H. Zaman. 2011. "Estimating the Short-Run Poverty Impacts of the 2010–11 Surge in Food Prices." Policy Research Working Paper 5633, World Bank, Washington, DC.

Jayne, T., A. Chapoto, and J. Govereh. 2007. "Grain Marketing Policy at the Crossroads: Challenges for Eastern and Southern Africa." Paper prepared for the FAO Workshop on "Staple Food Trade and Market Policy Options for Promoting Development in Eastern and Southern Africa," March 1–2, Rome.

Kshirsagar, V., and J. Baffes. 2011. "The Nature and Structure of Global Food Markets." World Bank, Washington, DC.

Laborde, D., W. Martin, and D. van der Mensbrugghe. 2011. "Implications of the Doha Market Access Proposals for Developing Countries." Policy Research Working Paper 5697, World Bank, Washington, DC.

Martin, W., and K. Anderson. 2011. "Export Restrictions and Price Insulation during Commodity Price Booms." Policy Research Paper 5645, World Bank, Washington, DC.

Mavroidis, P. 2007. *Trade in Goods.* New York: Oxford University Press.

Nijhoff, J., T. Jayne, B. Mwiinga, and J. Shaffer. 2002. "Markets Need Predictable Government Actions to Function Effectively: The Case of Importing Maize in Times of Deficit." Policy Synthesis 6, Food Security Research Project, Lusaka.

Nijhoff, J., D. Tschirley, T. Jayne, G. Tembo, P. Arlindo, B. Mwiinga, J. Shaffer, M. Weber, C. Donovan, and D. Boughton. 2003. "Coordination for Long-Term Food Security by Government, Private Sector and Donors: Issues and Challenges." Policy Synthesis 65, Michigan State University.

Nyameino, D., B. Kagira, and S. Njukia. 2003. "Maize Market Assessment and Baseline Study for Kenya." Regional Agricultural Trade Expansion Support Program, Nairobi.

OECD (Organisation for Economic Co-operation and Development). 2011. "Agricultural Policy Monitoring and Evaluation 2011: OECD Countries and Emerging Economies." Paris.

Rubey, L. 2005. "Malawi's Food Crisis: Causes and Solutions." Report for USAID, Lilongwe.

Saez, S. 2011. "Trade Policy Options and International Food Prices." World Bank, Washington, DC.

Shepherd, B. 2011. "Trade Costs in the Maghreb 2000–2009." Developing Trade Consultants Ltd.

Tschirley, D., D. Abdula, and M. Weber. 2005. "Toward Improved Marketing and Trade Policies to Promote Household Food Security in Central and Southern Mozambique." Michigan State University.

World Bank. 2005. "Infraestructura logística en Colombia." Report 35061-CO, Departamento de Finanzas, Sector Privado e Infraestructura, Región de América Latina y el Caribe, Washington, DC.

———. 2009a. "Eastern Africa: A Study of the Regional Maize Market and Marketing Costs." Agriculture and Rural Development Unit, Africa Region, World Bank, Washington, DC.

———. 2009b. "Rising Food Prices: Policy Options and World Bank Response." World Bank, Washington, DC.

———. 2011a. "Food Price Watch February 2011." Poverty Reduction and Equity Group, World Bank, Washington, DC.

———. 2011b. "Food Price Watch April 2011." Poverty Reduction and Equity Group, World Bank, Washington, DC.

———. 2011d. "Russia and Central Asia: Win-Win Approaches in Trade Integration." ECSPE and the Eurasian Development Bank Center for Integration, Washington, DC.

———. 2012. "De-Fragmenting Africa: Deepening Regional Trade Integration in Goods and Services." World Bank, Washington, DC.

WTO (World Trade Organization). 2011a. "Aid for Trade Work Programme 2012–2013—Deepening Coherence." JOB/DEV/12, Committee of Trade and Development, Geneva.

———. 2011b. *Report on G-20 Trade Measures.* Geneva.

5

Aid and International Financial Institutions

Summary and Main Messages

Official development assistance (ODA) has strengthened remarkably over the past decade, despite the disruptions of the global financial crisis centered in high-income countries. Net ODA reported to the Development Assistance Committee (DAC) of the Organisation for Economic Co-operation and Development (OECD) rose from 0.22 percent of donors' gross national income (GNI) in 2000 to 0.32 percent in 2010 and reached a record high of $127.3 billion in 2010 (at 2009 prices), very close to the target set at the Group of Eight Gleneagles Summit in 2005. And among the 15 European Union (EU) member countries that committed to raising ODA to 0.51 percent of GNI by 2010, 8 countries reached the goal and another 4 countries made significant progress toward it. There is some evidence that international coordination, notably the commitments made at Gleneagles, contributed to the rise in aid disbursements (Kharas 2010). Nevertheless, aid remains well short of the goal of 0.7 percent of GNI set by the United Nations some 40 years ago and substantially below various estimates (Atisophon and others 2011) of annual disbursements required to meet the Millennium Development Goals (MDGs). Further, a key concern is that it

may take several years before the full impact of the global financial crisis on aid flows becomes apparent. This is underscored by the just-released (April 2012) preliminary OECD data indicating that ODA disbursements declined by 2.7 percent in 2011 (at 2010 prices), as fiscal consolidation in several DAC countries has cut into their aid budgets.

Despite the recent spikes in food prices, ODA commitments to agriculture, food, and nutrition are limited. The share of aid commitments directed toward agriculture, food, and nutrition has remained at about 10 percent since the MDGs were agreed upon in 2000. Further increases in aid for nutrition are particularly important: assistance for nutrition represents only 3 percent of total aid flows to agriculture, food, and nutrition, yet improved nutrition and gains in early childhood development are critical to economic progress.

Tight budget constraints in many donor countries underscore the need for improving aid effectiveness to meet the MDGs in 2015. Progress in improving aid effectiveness has fallen short. Only 1 of the 13 global targets set out in the Paris Declaration on Aid Effectiveness (2005) to be achieved by 2010 has been met, and only limited progress has been achieved on the other 12. Directing a larger share of disbursements to country programmable aid (CPA, a core subset of ODA, which accounts for about 60 percent of total DAC gross bilateral ODA and excludes unpredictable components such as food aid and aid flows that do not have direct development impacts such as donor administrative costs) would also help to mitigate the impact of weakened aid flows.

The very welcome expansion of new donors has raised new challenges for aid recipients and has led to shifts in the international aid agenda. While data remain limited, the Bank estimates that aid disbursements by non-DAC bilateral donors (including new donor middle-income countries) and private actors such as philanthropic organizations reached $63.5 billion in 2009. The lion's share, $52.5 billion, came from private nongovernmental organizations (NGOs) (Hudson Institute 2011) and the remaining $11 billion came from non-DAC donors (accounting for $7.3 billion) and new middle-income donors Brazil, China, India, the Russian Federation, and South Africa (together accounting for $3.7 billion) (Zimmermann and Smith 2011). The rapid rise in the number of donors and projects has increased the administrative burden facing recipients, particularly for fragile and conflict-affected states (OECD 2011c). The sharp rise in stakeholders has contributed to important shifts in the aid agenda, including calls for strengthening country leadership and ownership of the aid management process; promoting a more inclusive process of development cooperation; improving delivery, measurement, and monitoring of results; and improving harmonization and transparency of aid management and delivery practices—common goals that the development community endorsed in the Global Partnership for Effective Development at the Fourth High Level Forum on Aid Effectiveness in Busan in 2011. Additionally, participation by new actors, particularly the private sector, has led to calls for greater emphasis on innovation (Gates 2011).

With increased international trade, foreign investment, and remittances flows, ODA is now viewed as only one component of many international activities that support development and poverty alleviation (Zoellick 2011). Nevertheless, ODA remains particularly important for low-income countries. It represented more than 60 percent of their external finance during 2005–10, compared with a mere 4 percent for middle-income countries (Adugna et al. 2011). ODA is critical for fragile and conflict-affected states, where integration with global markets has been severely hampered. Recognizing that few conflict-affected countries would achieve a single MDG by 2015, the New Deal for Engagement in Fragile States (which stakeholders endorsed at the Fourth High Level Forum in Busan) sets out 5 priorities to work toward: legitimate politics, justice, security, economic foundations, and revenues and services.

Fiscal consolidation in many large donor countries is likely to slow the growth of aid in coming years. Donor reports indicate that the growth of disbursements of country programmable aid could fall from an average of 5 percent a year recorded during 2001–10 to an average of 2 percent during 2011–13. This implies an annual per capita decline of 0.2 percent of CPA disbursements for recipient countries. Disbursements to countries in conflict or fragile situations may decline by an annual 2.1 percent on a per capita basis, although they would remain four times the per capita level expected for other aid recipients. If realized, lower per capita CPA disbursements could have significantly negative fiscal implications for the countries affected—and potentially for the achievement of the MDGs. This potential aid decline underscores calls at the Fourth High Level Forum on Aid Effectiveness in Busan to focus on results, to scale up aid for fragile states and underaided countries, and to improve aid coordination.

Recent trends in the disbursement and composition of aid

Official development assistance strengthened substantially in 2010, despite ongoing challenges tied to the global financial crisis and limited fiscal space in many high-income countries. DAC member countries' bilateral ODA net disbursements increased by 6.3 percent in constant dollars to $127.3 billion, the highest level on record, exceeding the previous peak of $122.3 billion in 2005. This increase followed weak volume growth of 1 percent in 2009, as the global economy grappled with recession. Bilateral ODA net disbursements rose to 0.32 percent of DAC donors' GNI in 2010, up from 0.22 percent in 2000 and the highest share since the record 0.33 percent posted in 2005 (figure 5.1). Of the $127 billion in ODA net disbursements from DAC countries, 29 percent was directed to low-income countries, 18 percent to middle-income countries, and 30 percent to multilateral institutions.[1]

In contrast to DAC bilateral ODA, multilateral net disbursements for development contracted by 1.6 percent in 2010 to $13.2 billion in constant 2009 prices. Since multilateral disbursements accounted for only 9 percent of total disbursements (DAC bilateral ODA and multilateral aid) in 2010, the rise in DAC bilateral ODA more than offset the multilateral decline; aggregate DAC bilateral and multilateral net aid disbursements reported to the OECD reached a record high of $147.5 billion in 2010 at constant 2009 prices.

The increase in ODA in 2010 continued the general trend of rising flows throughout much of the decade. DAC bilateral ODA registered a cumulative net gain over the decade of nearly $48 billion in constant prices. This 60 percent real increase was by far the largest decadal gain since data collection began in 1960 (table 5.1). In the 40 years through 2000, DAC ODA grew by an annual average of 2.1 percent in real terms, while during the decade through 2010 the pace accelerated to an average 5.5 percent. The general trend of rising annual flows during the 2000s was only briefly constrained by the onset of the global financial crisis in 2008, with a sharp deceleration in growth to 1 percent in 2009.

To a large extent the buoyancy in aid disbursements over the decade is tied to the 31.6 percent surge (in real terms) in DAC bilateral ODA in 2005 that is associated with international agreements that targeted

FIGURE 5.1   DAC members' net ODA bilateral disbursements

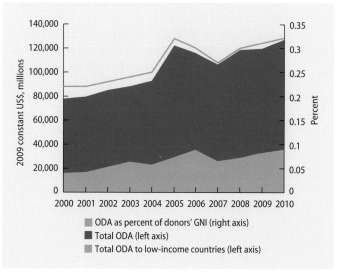

Source: OECD DAC.

TABLE 5.1   Decadal changes in bilateral official development assistance

Constant 2009 prices, decade ending in year noted

Bilateral ODA	1970	1980	1990	2000	2010
Percent growth	0.8	46.5	38.2	−9.2	60.0
Level change (US$ millions)	339	19,418	22,870	(7,902)	47,799

Source: OECD DAC.

substantial increases in international aid and debt relief. These include agreements reached in 2002 in Monterrey and in 2005 in Gleneagles and Paris and the Multilateral Debt Relief Initiative (MDRI), as well as the Heavily Indebted Poor Countries (HIPC) Initiative dating back to 1996. The surge in ODA in 2005 in particular reflects debt relief tied to HIPC, MDRI, and traditional debt relief mechanisms under the Paris Club, which in aggregate accounted for 17 percentage points of the 31.6 percent real increase in ODA for the year. Debt relief represented 10 percent of ODA over the decade (2001–10), peaking at 22.2 percent in 2005. This level compares with a somewhat smaller average share of 7.8 percent during the 1990s and a much smaller 2.3 percent share in the three earlier decades (1960s through 1980s). The upswing in multilateral ODA helped reverse the contractions in DAC bilateral ODA of 5.2 percent in 2006 and 8.1 percent in 2007 (in real terms).

2010 was the deadline to achieve very ambitious targets that donors and partner countries set for themselves in 2005 to increase development aid flows in an effort to help realize achievement of the Millennium Development Goals in 2015. More specifically, at the G-8 Gleneagles Summit in 2005, donors agreed to raise annual ODA disbursements by about $50 billion by 2010 and the 15 EU countries that are members of OECD DAC committed to raise ODA flows to 0.51 percent as a share of GNI by 2010. While neither target was met, significant progress was made despite the severe disruptions tied to the global financial crisis since 2008—and the commitments made at Gleneagles and other international initiatives (such as the High Level Forums) appear to have contributed to the rise in aid disbursements (Kharas 2010). Aid disbursements reached a record high of $127.3 billion in 2010 and helped bring donors very close to achieving the G-8 Gleneagles target for 2010 of $130 billion (at 2009 prices). Additionally, 8 of the 15 EU member countries that committed to the 0.51 percent target reached it, and 4 countries made significant progress toward the goal.[2]

More countries made larger ODA disbursements relative to their GNI over the past decade than during the 1990s—or indeed since the 1960s. The individual country efforts of smaller countries have exceeded those of the larger DAC bilateral donor countries. EU member countries have led this trend. For example, ODA disbursements rose by a minimum of one-tenth of a percentage point as a share of GNI in Belgium (from 0.53 percent to 0.64 percent), Denmark (0.81 percent to 0.91 percent), Ireland (0.42 percent to 0.52 percent), Luxembourg (0.79 percent to 1.05 percent), Spain (0.27 percent to 0.43 percent), and the United Kingdom (0.47 percent to 0.57 percent) between 2005 and 2010, reflecting a concerted effort to meet their Gleneagles 2010 commitment. By contrast, U.S. ODA disbursements averaged 0.17 percent of GNI during the decade through 2010, (notably rising from an average of 0.13

percent in the 1990s). As a consequence, the simple average of the DAC country effort has come to exceed, by a wide and growing margin, the income-weighted average (which weighs the given country's effort by its GDP, and thus giving larger economies a larger weight). While the DAC countries' total weighted ODA disbursements as a share of GNI rose to 0.32 percent in 2010, the average unweighted country share rose to a record high of 0.47 percent in 2010 (up from 0.36 percent in 2000) (figure 5.2).

The considerable rise in aid flows over the past decade has been accompanied by a significant reorientation of flows toward low-income countries, where aid also represents a much larger source of external financing needs (figure 5.3). Low-income countries accounted for a peak 61.9 percent of aid flows in 2010, compared with 46.9 percent and 44.3 percent in 2000 and 1990, respectively.[3] In contrast, ODA flows to middle-income countries fell from 56.7 percent of the total in 1990 to 38.1 percent in 2010. The rise in the low-income-country share of ODA flows represents a recent acceleration of a long-term trend: these countries received only 35 percent of ODA flows to developing countries in the 1960s (and a historical low of 27 percent in 1961, with available data starting in 1960). The recent rise in disbursements to them in part reflects efforts tied to the war on terrorism. Nevertheless, if total ODA disbursements

FIGURE 5.2 DAC ODA as a share of donor GNI
Percent share, current prices

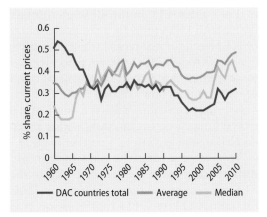

Sources: OECD Creditor Reporting System and World Bank.

from all donors to Afghanistan are excluded, the trend is still evident and becomes more pronounced later in the 2000s. And notably, the surge in flows to middle-income countries during the mid-2000s is accounted for by disbursements to Iraq (largely in the form of debt forgiveness). Aid flows are a significant source of external financing for low-income countries, with ODA representing more than 60 percent of total external financing for them from 2005 to 2010, in contrast to a mere 4 percent for middle-income countries, where foreign direct investment (FDI) and other sources of private financing accounted for more than three-fifths of external financing needs (Adugna et al. 2011).

Regional shifts in ODA reflect the reorientation in aid toward low-income countries and an increased concentration of flows by large donor countries toward strategically important recipient countries. A key example of the latter is the United States, which has concentrated its bilateral aid flows in Afghanistan over the past decade (along with Iraq, a lower-middle income country; see figure 5.3).

In **South Asia**, real ODA disbursements to Afghanistan increased from $1.6 billion

in the 1990s to $27.9 billion in the 2000s, whereas Bangladesh and India experienced a decrease in real ODA disbursements of about 20 percent. Afghanistan accounts for 41 percent of real ODA to South Asia, followed by Pakistan (17 percent), India (16 percent), and Bangladesh (12 percent).

In the **Middle East and North Africa**, the Arab Republic of Egypt received more than 50 percent of regional ODA disbursements from 1990 to 1999, followed by Morocco (11 percent) and Jordan (8 percent). That changed during the 2000s, when disbursements to Iraq surged as it became a strategic focus for the United States. Iraq has received more than $60 billion since 2000, or 59 percent of regional ODA flows during the 2000s (in real terms).

Among the other developing regions, **Sub-Saharan Africa** also saw a significant upswing in real ODA disbursements during 2001–10 compared with 1991–2000, reflecting efforts by donors to support acceleration in progress toward meeting the MDGs. Nigeria, Democratic Republic of Congo, and Tanzania experienced the largest increases in aid disbursements

FIGURE 5.3   Net ODA disbursements to low- and middle-income countries and by region

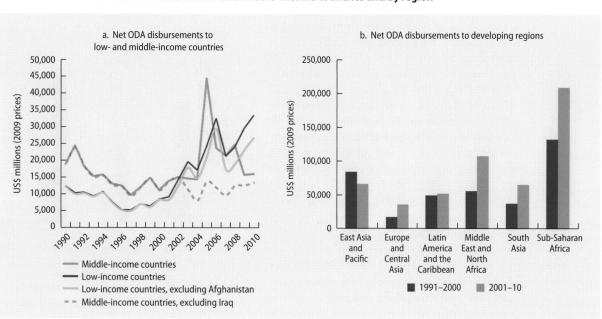

Source: OECD DAC.

to the region, accounting for 25 percent of regional ODA disbursements. **Europe and Central Asia** also saw a substantial increase in ODA, albeit from a low base. Real aid disbursements shifted from Bosnia-Herzegovina (30 percent of regional disbursements) and Turkey (17 percent) toward Serbia. Real ODA to Serbia increased from $1.6 billion in the 1990s to $10.3 billion in the 2000s.

Regional ODA flows have remained roughly stable in **Latin America and the Caribbean** in real terms. Colombia saw the largest percentage rise in disbursements, from $2.1 billion in the 1990s to $7.8 billion in the last decade (a real increase of 262 percent). In contrast to the rest of the developing regions, ODA disbursements in **East Asia and Pacific** declined markedly in real terms in the last decade, as the region made strong gains toward poverty alleviation. In particular, the aid decline reflects a fall-off in flows of more than 30 percent to the large regional economies of China, Indonesia, and the Philippines. These declines more than offset the 159 percent increase in disbursements to Vietnam, which became the top regional recipient with 22 percent of the region's total ODA in 2010.

Another important development over recent years, attendant with increased focus of aid flows to low-income countries, is that aid is increasingly being directed to fragile states and situations (FSS).[4] The severity of the situations in FSSs has widespread effects that are manifest locally, regionally, and globally. The 32 countries categorized as FSSs (according to the International Development Association, or IDA) accounted for about 18 percent of total net bilateral ODA disbursements and multilateral development assistance and 25 percent of net bilateral disbursements from DAC countries in 2010. These countries represent 425 million people. Some other definitions of countries in fragile situations include countries with as many as 1.5 billion people (World Bank 2011b). ODA disbursements to fragile states increased in 2001–10 (both including and excluding Iraq and Afghanistan), while aid to nonfragile

states held steady (both including and excluding China and India).[5]

Corresponding to the trend of higher ODA to low-income countries, the level of net ODA received on a per capita basis has shifted increasingly toward countries that are furthest from achieving the MDGs. For example, the group of countries that have met or are currently on track to achieve no more than two MDGs received an annual average of $48 per capita in 2008–10, up by 20 percent in real terms compared with 1990–92, and more strikingly up 85 percent compared with 2000–2002 (figure 5.4). This rise in flows to countries that are furthest from attaining the MDGs represents an important trade-off between need and performance, because aid effectiveness (improvement in outcomes per dollar spent) in these countries is likely to be weaker compared with other countries closer to the 2015 targets. However, successfully tackling circumstances where performance has been severely hampered (by conflict or natural disasters, for example) also provides scope for the greatest possible gains.

FIGURE 5.4 Net ODA received per capita by groups of countries ranked by MDG targets met or on track to be met by 2015

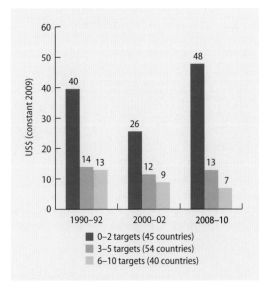

Source: World Development Indicators database, OECD DAC, and World Bank staff calculations.

Recent trends in the composition of aid for agriculture, food, and nutrition

Despite the spike in food prices, ODA commitments from all donors to agriculture, food, and nutrition did not increase as a share of total ODA between 2000 and 2010. While aid commitments from DAC bilateral ODA and multilateral developmental assistance to agriculture, food, and nutrition rose from $8.7 billion in constant terms in 2000 to near $16 billion in 2010, the share remained roughly unchanged at close to 10 percent. In the mid-2000s, increased focus was paid to debt forgiveness (particularly for highly indebted poor countries and Iraq). As a result, and despite a 75 percent increase in committed support to agriculture, food, and nutrition from all donors, the actual share in total ODA commitments declined from 10 percent in 2000 to about 7 percent in 2006. Excluding debt forgiveness, ODA for agriculture, food, and nutrition from all donors has remained more stable since the mid-2000s at about 10 percent of total remaining commitments, 1 percentage point below the 11

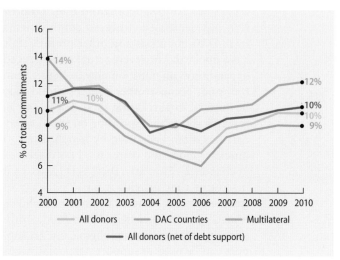

FIGURE 5.5   **Share of committed ODA to food, nutrition, and agriculture by donor**

Source: World Bank staff calculations based on OECD DAC.

percent share recorded during the first few years of the decade (figure 5.5).

Assistance for nutrition represents only 3 percent of total agriculture, food, and nutrition commitments, despite widespread evidence that improved nutrition and gains in

FIGURE 5.6   **Composition of committed ODA and commitments by donors in year 2010**
Constant 2009 millions, unless otherwise noted

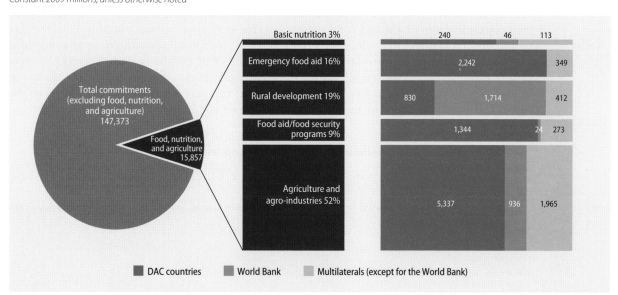

Source: World Bank staff calculations based on OECD DAC.

early childhood development are key in making long-term progress in development (figure 5.6). Since 2000 support from IDA to nutrition has decreased, whereas commitments from DAC countries and other multilaterals have doubled. However, actual aid-financed expenditures on nutrition may be higher than reported, because other sources of aid may be devoted to purchasing food. For example, research shows that spending on social safety nets has often been used by beneficiaries to purchase more and better food (as discussed in chapter 2). Similarly, there is evidence that programs that provide a basic package of free health care services to poor households is also spent by beneficiaries on food. Additionally, aid delivered as fungible budget support can be used to support particular needs (which may be nutrition) or sectors of the economy (which may be agriculture).

More than 40 percent of food-related development assistance commitments were directed to agriculture and agro-business in the year 2000. Remaining aid commitments were intended for programs related to food aid and food security (30 percent), rural development (16 percent), emergency food aid (7 percent), and basic nutrition (2 percent). Recent data (2010) show that agriculture and agro-industries, rural development, and emergency food aid have gained significantly in aid importance, whereas committed resources to food aid and food security programs have considerably decreased (table 5.2). This pattern illustrates a shift in the donor community to focus on alleviating the short-term impact of food crises on the most vulnerable, while at the same time providing support to programs aimed at bolstering productivity and long-term growth in agriculture. In this new architecture, DAC countries have concentrated their efforts on emergency response and food aid programs, whereas international financial institutions (IFIs), particularly the World Bank, have focused on rural development, agriculture, and agro-industries. (See the annex for further discussion of the IFIs' response to the recent spikes in food prices.)

ODA commitments by income group for agriculture, food, and nutrition have increasingly shifted toward low-income countries. On average during the decade through 2010, this group received about two-thirds of total ODA commitments for this category. ODA commitments for basic nutrition for low-income countries accounted for 0.2 percent of total ODA commitments for all categories during the decade, twice the amount received by lower-middle-income countries (0.1 percent), and twice again that received by upper-middle-income countries (0.05 percent) (see figure 5.7 and the appendix for the classification of economies).

Expansion of the donor community

Recent years have witnessed an expansion and diversification of the donor base for concessional aid, notably from NGOs and, to a lesser extent, the emergence of a number of middle-income countries as new donors (even while in some cases they are still receiving ODA). This expansion of the donor community appears to be reinforcing

TABLE 5.2   Composition of committed ODA to nutrition, food, and agriculture

Category	Creditor Reporting System code	2000		2010	
		Constant 2009 $	Share of commitments (%)	Constant 2009 $	Share of commitments (%)
Basic nutrition	12240	216	2	398	3
Emergency food aid	72040	603	7	2,598	16
Rural development	43040	1,358	16	2,960	19
Food aid/food security programmes	52010	2,640	30	1,644	10
Agriculture and agro-industries	3110-95	3,863	45	8,257	52
Total		8,680	100	15,857	100

Source: OECD DAC, Creditor Reporting System.

FIGURE 5.7   ODA commitments by income group

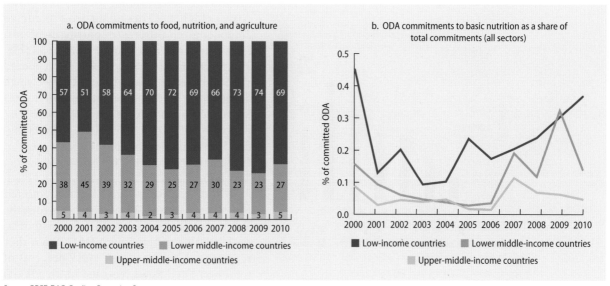

Source: OECD DAC, Creditor Reporting System.
Note: See the appendix for the classification of economies.

the increased concentration of aid flows to low-income countries (noted earlier). While the broadening of the donor base has been apparent for decades, it intensified in the 1990s and particularly in the second half of the 2000s, and in part simply reflects better reporting of aid flows. The proliferation and increased diversity of donors bring a number of benefits aside from increased aid disbursements—including complementarities, additional resources, and technical expertise—but the proliferation of donors also poses important new challenges, including rising transaction and administrative costs for both donors and recipients.

Data regarding concessional flows for development from NGOs, middle-income countries, and other newer donors remain extremely sparse, although they have improved. For example, the Gates Foundation has begun reporting aid disbursements to OECD DAC. Latest available estimates from the Hudson Institute indicate that private NGOs (foundations, philanthropist organizations, and corporations) provided $52.5 billion of international developmental flows in 2009 (latest available). Measured as a share of total bilateral ODA reported by OECD DAC, NGO contributions surged to

44 percent in 2009, up from zero reported aid in 1992.

More and more countries are providing ODA, and more and more countries are reporting data on their ODA disbursements. For example, the OECD reports on non-DAC ODA—aid flows from countries that are not members of the Development Assistance Committee—but not in the same detail as the DAC member countries provide. Twenty non-DAC countries reported to the OECD in 2009, up from 10 in 2000. The disbursement of non-DAC aid reported to the OECD increased to $7.3 billion in 2009 from $1.3 billion in 2000 (at constant 2009 prices). As a share of DAC bilateral ODA, non-DAC ODA rose to 6.1 percent in 2009 from 1.7 percent in 2000 in constant prices. The non-DAC countries include both high- and middle-income countries. The Arabian countries of Saudi Arabia, Kuwait, and the United Arab Emirates, and to a lesser extent, new European Union member countries Estonia, Latvia, Lithuania, Romania, and Slovenia, account for much of the increase in non-DAC flows reported to the OECD. Among the non-DAC middle-income countries, the Republic of Korea and Turkey markedly increased aid assistance during the decade, from near zero;

FIGURE 5.8   ODA from Brazil, Russia, India, China, and South Africa

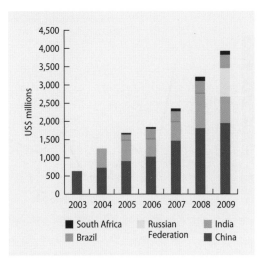

Sources: OECD DAC; Zimmermann and Smith 2011.

both were among aid recipients until a few years ago. Notably, Korea, an aid success story, became a DAC donor in 2009.

Data reporting has also improved in some middle-income countries that are new donors, although they do not report to the OECD.

Among those reporting official bilateral data are Brazil, China, India, Russia, and South Africa (BRICS). These data are difficult to compile, given different reporting methods. However, various estimates are available. Zimmermann and Smith (2011) estimated the South-South flows of the BRICS—which account for much of the increase in South-South aid flows in recent years—grew to $3.7 billion in 2009 from $0.6 billion in 2003 (figure 5.8). China accounted for 53 percent of the total ODA from BRICS in 2009 (reaching $1.9 billion up from an estimated $0.6 billion in 2003), and Russia and India accounted for 21 percent and 13 percent, respectively. While the aid flows from BRICS remain relatively small compared with DAC flows, they represented just over half (50.5 percent) of total non-DAC flows reported to the OECD in 2009, up from about one-fifth in 2003. To the extent data on aid from BRICS is available, given irregular reporting and different methodologies, some general trends are emerging. BRICS' aid is generally delivered to bilateral partners with a combination of conditional and nonconditional financing, usually without policy conditions, and is

TABLE 5.3   Key characteristics of BRIC financing

Characteristic	Brazil	Russian Federation	India	China
Key agency	Brazilian Cooperation Agency	Department of International Finance	Indian International Development Cooperation Agency[a]	No development agency (discussions are ongoing)
Key ministry	Ministry of External Relations	Ministry of Finance and Ministry of Foreign Affairs	Ministry of External Affairs	Department of Aid in Ministry of Commerce
Form	Loans and grants	Mostly grants and debt relief	Grants, credit lines, interest-free loans, and other concessional and nonconcessional loans	Grants, credit lines, interest-free-loans, and other concessional and nonconcessional loans
Country focus	Latin America and Africa (especially Lusophone)	Mostly Commonwealth of Independent States (especially Kazakhstan and the Kyrgyz Republic)	Neighboring countries (Afghanistan, Bhutan, Myanmar, and Nepal) and Africa	Widespread through large amounts concentrated in a small number of countries
Sector	Mostly agriculture, education, and health	Mostly general budget support	Grants for rural development, education, health, technical cooperation; loans for infrastructure and disaster relief	Mostly energy, transport, and communications, but also construction of schools and hospitals and prestige projects (such as stadiums)

Source: Mwase and Yang 2012.
Note: There is no systematic reporting of aid data for South Africa (Adugna et al. 2011).
a. Proposed in 2007 but not yet established.

often directed toward infrastructure and productive sector investment projects (table 5.3). Geographically, the BRICS tend to extend aid to neighboring countries, with the exception of China, which delivers significant aid flows to other regions. For example, India's aid is largely directed toward Afghanistan, Bhutan, Myanmar, and Nepal (Mwase and Yang 2012).

NGO concessional aid flows were the leading dynamic behind the doubling of real aid flows from 1992 to 2009. An aggregation of the Hudson Institute's NGO aid estimates with the OECD's bilateral DAC and non-DAC ODA disbursements—along with Zimmerman and Smith's (2011) estimates of ODA from BRICS—indicates that total global aid reached $183.3 billion dollars in 2009, up from $90 billion in 1992 (in real terms) (figure 5.9). NGOs represented 29 percent of this global total. OECD non-DAC countries and BRICS represented 6 percent—which together with NGOs accounted for more than one-third of reported total global aid in 2009. The Hudson Institute's estimates indicate that private NGO aid flows have come to eclipse ODA originating from new donor countries, despite their rapid growth over the last decade.

With the growing number of countries and organizations contributing ODA and development aid, the number of counterparts with which a recipient country engages has

increased markedly over the past decades. For instance, the OECD alone reports that in 2009 the average OECD donor was present in 71 of 152 ODA-eligible countries (73 for DAC countries and 69 for multilateral agencies), and that the average number of donors present in each recipient country was 21 (OECD 2011d). Aid fragmentation is greater if new agents (NGOs and middle-income countries) are included. As a consequence, partner country institutions are facing rising transaction costs, as they are required to dedicate more and more resources toward engagement with donor agents, while the average size of projects has declined (in part the result of improved reporting). Because anecdotal evidence suggests a growing duplication of effort, increased coordination among donors, as well as between donors and recipients, could generate significant gains in efficiency. The OECD reports that the global fragmentation ratio (number of nonsignificant donors compared with the overall number of donors) has risen in recent years, with fragile and conflict-affected states seeing the largest increase (table 5.4). As of 2009, two of every five DAC countries' aid relations were classified as nonsignificant, representing a total of about $2.9 billion or a mere 3 percent of total global CPA transactions. In response to the increased fragmentation of aid, several donors have undertaken efforts to concentrate aid disbursements on fewer recipient

FIGURE 5.9   Changes in sources of estimated global concessional developmental flows

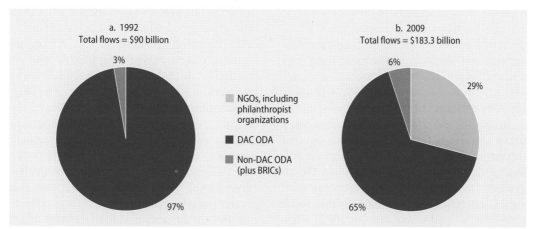

Sources: Hudson Institute 2010, 2011; OECD DAC; Zimmermann and Smith 2011; Fengler and Kharas 2010.

TABLE 5.4   **Aid fragmentation by income group and fragile and conflict-affected states**

Income group	Number of countries	Significant relations (A)	Nonsignificant relations (B)	Total relations (A+B)	Fragmentation ratio (F-ratio) B/(A+B) (%)	2008 F-ratio (%)	2004 F-ratio (%)
Lower	61	985	557	1,542	36	34	33
Lower-middle	48	590	531	1,121	47	46	46
Upper-middle	43	390	204	594	34	35	33
Total	**152**	**1,965**	**1,292**	**3,257**	**40**	**38**	**38**
Memo: fragile and conflict-affected	41	622	436	1,058	41	39	36
Memo: Other	111	1,343	856	2,199	39	38	38

Source: OECD 2011d.

countries. Notably, phasing out nonsignificant relations is largely an uncoordinated exercise and points to risks for recipient countries that have a high fragmentation ratio, such as fragile and conflict-affected states.

Aid effectiveness agenda

The marked changes in aid architecture over the past decade have coincided with a reexamination of ODA, with heightened scrutiny on the effectiveness of aid and results. A number of fundamental shifts in the aid agenda and approaches have become increasingly apparent. In part, new donors and nonstate participants have made demands for increased accountability and the effective use of money spent on development assistance. These demands have translated into calls for greater transparency in aid flows at all levels and across agents, and have highlighted the need for strengthening institutions to make them more results oriented for better monitoring of development programs and to broaden participation. In response, the development community has increasingly pursued various avenues to strengthen accountability and transparency among donors and recipients to improve outcomes, along with more rigorous measures of aid effectiveness.

Improving aid effectiveness has been a key focus at various international development forums over the past decade, especially following the signing of the Millennium Development Declaration in 2000. In particular, the aid effectiveness movement, with a focus on results, gained momentum in 2002, when

the Monterrey Consensus was adopted by more than 50 heads of state and gained the support of the International Monetary Fund, the World Bank, and the World Trade Organization. The Monterrey Consensus highlights the importance of development cooperation, recognizing that both domestic and international resources need to be mobilized for development. Donors agreed to ramp up aid flows. Participants at Monterrey also recognized that aid needs to be optimally used to accomplish the MDGs by 2015.

Since adoption of the Monterrey Consensus, the international framework for action on aid effectiveness has come to be articulated at the High Level Forums.[6] In 2003 the international aid community met in Rome for the First High Level Forum, which focused on harmonization among donors. Among the outcomes in Rome, donor institutions committed to improve coordination of their programs and to streamline activities.

At the Second High Level Forum in 2005 in Paris, stakeholders endorsed the Paris Declaration on Aid Effectiveness, an effort to comprehensively revamp the way donor and recipient countries work together to improve poverty reduction outcomes and achieve long-term sustainable development. The Paris Declaration on Aid Effectiveness placed the country ownership of policies and programs at the center of an international reform agenda to make aid more effective, with the international community recognizing that far-reaching and monitorable actions and greater focus on results would be necessary to improve the delivery and management of aid to maximize its contribution to the achievement of the MDGs. Five shared principles of aid effectiveness were set out in the Paris Declaration, along with more than 50 commitments. A distinct feature was the commitment by donors and developing countries to hold each other accountable for implementing the declaration through a set of clear indicators of progress with 13 measurable targets.[7] The 5 principles call for strengthened partner country ownership (partners setting the agenda); improved donor alignment with partners' agenda; harmonization across donors (establishing common arrangements, simplifying procedures, sharing information); managing for development results; and mutual accountability.

Participants recognized that while the recent proliferation of donor agents contributed to higher aid flows, it also led to fragmentation of aid, making it less predictable,

transparent and more volatile. Insufficient transparency was identified as an important bottleneck to improving outcomes,[8] including the need to achieve greater partner participation and harmonization with local objectives. Separately, a number of NGO initiatives have been undertaken to improve the tracking of aid flows and transparency along with the ability to assess aid effectiveness. Initiatives include those by Publish What You Fund (PWYF Index), the joint effort by the Center for Global Development and Brookings (Quality of Official Development Assistance, or QuODA), and Give Directly (box 5.1).

In 2008 the Third High Level Forum in Accra recognized an increased role for a range of development actors beyond the state that broadened the principle of ownership, strengthened participation of partner countries and other stakeholders (such as philanthropic foundations and global programs),

---

BOX 5.1   Examples of independent initiatives to improve aid effectiveness

Publish What You Fund (PWYF) campaigns for improved aid transparency, that is, more and better information about aid. The organization's 2010 Aid Transparency Assessment (the first global assessment for aid transparency) and its 2011 Index show that the aid information currently made available by donors is very limited and that there is a lack of comparable and primary data available. Their index compares transparency of 30 major donors by seven weighted indicators that fall into three categories— high-level commitment to transparency, transparency to recipient government, and transparency to civil society. PWYF reports wide variation in levels of donor aid transparency and significant weaknesses in many donors across the seven indicators.

The **Center for Global Development and Brookings** released the **Quality of Official Development Assistance** (QuODA) assessment in 2010. The QuODA assessment is similar to the PWYF's aid transparency index; however, it also seeks to address the broader issues of aid effectiveness and capture donor adherence to international standards outlined

in the Paris Declaration and Accra Agenda for Action. The QuODA assessments are compiled by constructing four dimensions or pillars of aid quality, built up from 30 separate indicators. The four dimensions are maximizing efficiency, promoting transparency and learning, fostering institutions, and reducing burden. Countries are ranked according to these assessments, which are intended to inform users of how much and what type of quality is "purchased" with the given country, agency, or multilateral aid delivery.

Give Directly is a nonprofit initiative to create an efficient and transparent way to provide aid. The organization allows individuals to donate money through its website to impoverished households in Kenya that Give Directly has identified. Give Directly transfers the donations electronically to the recipient's mobile phone, and the poor choose to what purpose they want to direct the funds.

Sources: http://www.Publishwhatyoufund.org/resources/ index/assessment; www.givedirectly.org; and Birdsall and Kharas 2010.

and deepened efforts to harmonize donor activities to improve the effectiveness of aid. Participants reaffirmed and deepened their commitments made in Rome and Paris and agreed on the need to accelerate progress toward improved cooperation and stronger results orientation. The Accra Agenda for Action consolidated the Paris Declaration principles and called for heightened focus on country ownership and leadership, more inclusive partnerships, and increased accountability for, and transparency about, development results. The Accra Agenda committed donors to publicly disclose regular, detailed, and timely information on volume, allocation, and, when available, results of development expenditure to enable more accurate budgeting, accounting, and auditing by developing countries. Additionally, recognizing the need to strengthen local capacity to monitor progress toward achieving the MDGs, international stakeholders at Accra mounted an effort to improve national statistical systems (box 5.2). The International Aid Transparency Initiative aims to help donor signatories meet this commitment in the most coherent and consistent ways, and to bring together donors, partner countries, civil society organizations, parliamentarians, and aid information experts to agree on common information standards applicable to aid flows.

In 2011 the international development community met at the Busan High Level Forum to assess progress on the MDGs and to determine where adjustments can be made to improve the outlook for meeting the goals by 2015, including ways to improve aid effectiveness and to better address the needs of under-aided countries and fragile states. Stakeholders reaffirmed the relevance of the aid effectiveness principles as stated in the Paris Declaration and deepened in the Accra Agenda for Action. In an evolving development landscape, stakeholders recognized that further efforts to increase the effectiveness of aid needed to be grounded in the broader development context, embracing the increasing diversity of development actors and seizing opportunities to leverage a wider range of sources of development finance. Discussions

involved a wider set of stakeholders than previous high-level forums, including civil society organizations, representatives of private sector organizations, and countries that had until Busan played a less active role in international dialogue on aid effectiveness (including a number of middle-income countries).

The Busan Partnership agreement is complemented by a range of initiatives presented as "Busan building blocks" that bring together groups of like-minded stakeholders around common goals and determined to take the agenda forward at the country level. These efforts were intended to operationalize principles and commitments set out in the outcome document, allowing for a deepening of commitments and, in places, further innovation on a voluntary basis at the country level. To support improved knowledge sharing and accountability, the Open Aid Partnership (OAP) was launched at Busan with support from the World Bank, the United Kingdom, Sweden, Spain, the Netherlands, Estonia, and Finland. More specifically, the OAP seeks to enhance transparency of public budgets, service delivery, and development assistance, which are critical for improving governance accountability and citizen engagement. It builds on the International Aid Transparency Initiative data standard to make aid information more accessible and meaningful to citizens, and is complementary to ongoing efforts of the Open Government Partnership. The Results and Accountability Building Block focused on operationalizing a transparent, country-led results framework and exploring additional initiatives at the country level aimed at improving the delivery, measurement, learning, and accountability for results.

Recognizing that few conflict-affected countries will achieve a single Millennium Development Goal by 2015, a number of countries and international organizations at Busan endorsed the New Deal for Engagement in Fragile States. The New Deal sets out 5 goals—legitimate politics, justice, security, economic foundations, and revenues and services—to give clarity on the priorities in fragile states.[9] Stakeholders agreed that the

BOX 5.2   Better statistics for all: Monitoring the millennium development goals

The need for reliable and timely statistics to monitor the results of development programs was recognized long before the Millennium Development Goals were promulgated, but the widespread attention given to their quantitative targets has increased the demand for regular and uniform reporting of key indicators. Faced with large gaps in the international database, the Partnership in Statistics for Development in the 21st Century (PARIS21) was established in 1999 to coordinate efforts to increase the statistical capacity of developing countries. In 2004 the Second Round-table on Managing for Development Results endorsed the Marrakech Action Plan for Statistics (MAPS), establishing an international agenda for support to statistics in developing countries. Subsequently the Accra Agenda for Action made broad commitments on behalf of donors and developing countries to strengthen national statistical systems. More recently, at the November 2011 Fourth High Level Forum on Aid Effectiveness, held in Busan, heads of state, ministers, and other representatives of developing and developed countries endorsed a global action plan for statistics.[a] This is the first time a statistical action plan has received explicit endorsement globally from the highest political levels.

Much progress has been made. The quality of statistics as measured by the World Bank's statistical capacity indicator has improved from its bench-mark level of 54 in 1999 to 67 in 2011 (see table). The availability of data for monitoring the MDGs has improved commensurately: in 2003 only 4 countries had two data points for 16 or more of 22 principle MDG indicators; by 2009 118 countries met this measure (OECD 2009b). Of 79 low-income IDA countries, only 8 do not have a national strategy for the development of statistics and are not planning to prepare one. Implementing these strategies is well under way in many countries. After the 2010 census round concludes in 2014, 98 percent of the world's population will have been counted. Since donors began reporting support for statistical capacity development in 2008, financial commitments to statistics increased by 60 percent to $1.6 billion over the period 2008–10. More than 55 developing countries have improved their practices in data collection, management, and dissemination of household surveys. The United Nations Interagency and Expert Group on the MDGs has conducted a series of regional workshops aimed at improving the monitoring of the MDGs and has reported annually on progress.

a. Busan Partnership for Effective Development Co-operation. Para 18.c of http://www.aid effectiveness.org/busanhlf4/images/stories/ hlf4/OUTCOME_ DOCUMENT_-_FINAL_EN.pdf.

World bank statistical capacity index of IDA-eligible countries

	All		Sub-Saharan Africa		non-Sub-Saharan Africa	
	1999	2011	1999	2011	1999	2011
Overall	54	67	49	58	49	68
Methodology	44	57	35	39	39	59
Source data	53	65	46	53	51	65
Periodicity	65	81	65	82	58	81

Source: For more on the World Bank statistical capacity index, see http://data.worldbank.org/data-catalog/bulletin-board-on-statistical-capacity
Note: Countries included are those IDA-eligible countries with a population above 1 million.

current ways of working in fragile states need serious improvement, and that despite the significant investment and the commitments of the Paris Declaration and the Accra Agenda, results and gains in value for money have been modest. Stakeholders also recognized that transitioning out of fragility is long, political work that requires country leadership and ownership. The New Deal recommends the use of peace-building and state-building goals as an important foundation to enable progress toward the MDGs. The New

Deal also calls for an increase in the predictability of aid, including by publishing three-to-five year indicative forward estimates (as committed in the Accra Agenda for Action).

Another main outcome from Busan was the creation of a global partnership to support global-level monitoring and accountability through a new and more inclusive development agenda. Delegates in Busan, including Brazil, China, and India, endorsed the Busan Partnership for Effective Development Cooperation (Global Partnership) on common goals, shared principles, and differentiated commitments.[10] The Global Partnership recognizes that, whereas the different types of aid donors should work toward common goals, donors can achieve them by "embracing their respective and different commitments." The new development agenda is based on 4 principles: ownership of development priorities by developing countries, focus on results, inclusive development partnerships, and transparency and accountability. Donors pledged to make their aid information available to the public and to help recipient countries establish transparent public financial management and aid information management systems. To increase focus on development results, the Global Partnership seeks to strengthen partner country ownership and to strengthen their policies and core institutions through the creation of transparent and country-led results

frameworks and platforms, and ensure that increased aid inflows are absorbed and spent efficiently to enhance growth. Donors pledged to ramp up efforts to fully unwind and end the practice of tied aid (which requires recipient countries to spend aid dollars on deliverables from companies in donor countries)—including efforts to improve the quality and transparency of reporting on the process of untying aid.

While remarkable progress has been made toward increasing aid disbursements, progress toward improving aid effectiveness has been less impressive—aside from achieving broad consensus in identifying specific areas that need to be addressed. In Paris in 2005, for example, the development community of donors and recipients agreed to pursue, and hold each other accountable for, reaching 13 very ambitious global targets. By the 2010 deadline only 1 of the targets had been met, although there were some apparent gains toward achieving the other targets (table 5.5). While many of the reforms that were needed to reach the Paris Declaration targets were widely recognized as being very ambitious, the targets are nevertheless attainable.

Some measurable, if limited, improvements have been made in aid effectiveness, particularly in recipient countries, and specifically in the areas of monitoring capacity and policy framework, as well as in collaboration and harmonization among donors

TABLE 5.5   Progress toward Paris Declaration targets
Survey outcomes in percentages, unless otherwise noted

Indicator	2010 Actual	2010 Target	Status
Operational development strategies	37	75	not met
Reliable public financial management (PFM) systems	38	50	not met
Aid flows aligned with national priorities	41	85	not met
Strengthen capacity by coordinated support	57	50	**met**
Use of country PFM systems	48	55	not met
Strengthen capacity by avoiding parallel projects (number)	1,158	565	not met
Aid is more predictable	43	71	not met
Aid is untied	86	> 89	not met
Use of common arrangements, procedures	45	66	not met
Joint missions	19	40	not met
Joint country analytic work	43	66	not met
Results-oriented frameworks	20	36	not met
Mutual accountability	38	100	not met

Source: OECD 2011a.

and partner countries. According to a progress report on implementing the Paris Declaration based on 2010 surveys of donors and recipients, the proportion of developing countries with sound national development strategies in place more than tripled from 2005 to 2010 (OECD 2011a). The results-oriented frameworks to deliver results and monitor progress against national development priorities are in place in one-fourth of reporting developing countries, and statistics related to the MDGs are becoming increasingly available.

Progress has been moderate or mixed in the areas of capacity development and the quality of country public financial management (PFM) systems in partner countries. The OECD also reports that support for capacity development is often supply driven, rather than geared toward the developing countries' needs, and remains an area for further improvement (OECD 2011a). Nevertheless, donors met the target on technical cooperation. And while more than one-third of partner countries showed improvement in the quality of PFM systems from 2005 through 2010, one-fourth experienced setbacks. Donor countries are using partner country PFM systems more extensively than they did in 2005, but they have fallen short of the target. More specifically, donors' use of country PFM systems could be strengthened where the systems have been made more reliable.

Some donors have made measurable progress and have introduced innovative approaches and reforms to improve aid effectiveness and to meet the Paris Declaration targets. Among bilateral donors for example, the United Kingdom's Department for International Development (DFID) initiated results-based financing in 2010, focusing specific outputs at a more micro level by offering incentives to a service provider or beneficiary of services (de Hennin and Rozema 2011). The intent is that from 2011 through 2014, all of DFID's bilateral ODA allocations will be based on evidence-supported "results offers," which are competitively bid upon by country and regional offices around the world (Birdsall 2010). Some IFIs, including the World Bank, have made significant gains across many of the Paris Declaration targets (box 5.3). Another example is the Consultative Group on International Agricultural Research (CGIAR), which has improved collaboration and harmonization to enhance results on the ground (box 5.4).

Multilateral development banks (MDBs) have also made significant progress in attaining the Paris Declaration targets. The country-led development model—direct funding of government expenditures, support for the private sector based on national priorities, or both—has allowed for the mainstreaming of aid effectiveness principles. As a whole, MDBs have outperformed development partner performance, according to the 2010 Paris Declaration survey (table 5.6). Key challenges, identified in the survey, remain, however. These include the use of country systems (especially procurement), aid predictability, and common arrangements with other development partners.

Areas of limited or no progress include untying aid, common arrangements or procedures, aid fragmentation, and medium-term predictability of aid. In its progress report, the OECD reports that the untying of aid shows no improvement and that aid is becoming increasingly fragmented (OECD 2011a). Survey results also show limited progress has been achieved among donors to implement common arrangements or procedures and to conduct join missions and analytic work. Donor information on future aid disbursements remains very limited, and hence the predictability of aid remains a key challenge for developing country governments. Additionally, the majority of partner countries have yet to implement thorough, mutual (government-donor) reviews of performance.

Country programmable aid and outlook for ODA flows through 2013

Among efforts to improve aid effectiveness, DAC donors in 2007 agreed to provide annual forward spending plans for country programmable aid (CPA)—to improve the

BOX 5.3   The World Bank has made significant progress on the aid effectiveness agenda, but there is room for improvement

Because of its mission, mandate, and country-driven business model, the World Bank demonstrates strong performance on the Paris Declaration monitoring survey, the main tool for tracking progress globally on the aid effectiveness agenda. In 2011 the Bank's results on the survey were better than the overall development partner average, and the Bank has met or is close to meeting the majority of targets (box table). The World Bank Group also performs well on other independent rankings such as those conducted by Publish What You Fund and the Center for Global Development and Brookings. (see box 5.1).

Paris Declaration survey indicators (development partner performance)	Overall results			World Bank results			
	Target (%)	2010[a] (%)	Meeting target?	Target (%)	2010[a] (%)	Meeting target?	Progress since 2005?
3. Aid flows are aligned with national priorities (aid on budget)	85	41	■	85	74	■	+
4. Strengthen capacity by coordinated support (technical assistance)	50	57	■	50	73	■	+
5a. Use of country PFM systems	55	48	■	51	69	■	+
5b. Use of country procurement systems	—	44	—	50[b]	54	■	+
6. Strengthen capacity by avoiding parallel project implementation units	−67	−32	■	−67	−80	■	+
7. Aid is more predictable	71	43	■	83	61	■	−
8. Aid is untied	89	86	■	100	100	■	=
9. Use of common arrangements or procedures (program based approach)	66	45	■	66	59	■	+
10a. Joint missions to the field	40	19	■	40	29	■	+
10b. Joint country analytic work	66	43	■	66	59	■	+

Source: OECD-DAC and World Bank.
■ denotes the target is achieved; ■ denotes the target is nearly achieved (gap is about 10%). ■ denotes the target is not achieved.
a. Indicators 3, 5a, 5b, 6, and 7 are calculated for the 30 countries that participated in the 2006 baseline survey and the 2011 survey.
b. The 2008 Accra Agenda for Action target of 50% is applied.
— = not available.

Nevertheless, while it has made significant gains across the Paris Declaration survey indicators, the Bank fell significantly short of the target in some areas, particularly in making aid more predictable and, to a lesser extent, in donor coordination (joint missions to the field and joint country analytic work) and in aligning aid flows with national priorities.

The Bank has been playing a key role in shaping the international aid effectiveness agenda over the years, and has mainstreamed the aid effectiveness agenda at the country and corporate levels. The World Bank's aid effectiveness priorities are based on a country-based business model and an ongoing work program focused on:

• *Country ownership and leadership*: Country-led aid management and coordination is paramount (evolution away from donor harmonization); use of country systems (procurement, financial management, safeguards, statistics, monitoring and evaluation, budget, project management) is critical to country ownership and leadership; and Capacity development is key to strengthening country systems and building effective institutions.

• *Development partnerships beyond aid*: New partnerships and approaches need to be recognized—DAC donors and traditional donor/recipient models of aid are no longer the only approach; the aid landscape is evolving—middle-income countries play an increasingly important role as providers of development assistance; foundations, global funds and programs, NGOs and the private sector are also major providers of assistance; partners use a multiplicity of approaches—South-South cooperation, knowledge exchange, technology transfer, foreign direct investment, trade, financing, aid.

BOX 5.3   The World Bank has made significant progress on the aid effectiveness agenda, but there is room for improvement (continued)

- *Transparency for Results*: The World Bank is a path-breaker on transparency—access to information policy, open data initiative, project database, international aid transparency initiative. The Bank is also strong on results—IDA and Corporate scorecard; core sector indicators country, project, and program level results frameworks; statistical and monitoring and evaluation capacity development; Development Impact Evaluation Initiative.

- *Fragile and Conflict Situations*: These countries are a special focus for the Bank, including supporting better aid management and coordination.

Source: World Bank. Food Crisis: Issue Briefs. Available online at World Bank. "The World Bank and Aid Effectiveness: Performance to Date and Agenda Ahead." November 2011.

BOX 5.4   CGIAR: Improved collaboration and harmonization to strengthen delivery

The Consultative Group on International Agricultural Research (CGIAR) is a global partnership that unites 15 International Research Centers and partner organizations engaged in research for sustainable development with the major global funders of this work. Its vision is to "reduce poverty and hunger, improve human health and nutrition, and enhance ecosystem resilience through high-quality international agricultural research, partnership and leadership." The funders include developing- and industrial-country governments, foundations, and international and regional organizations.

The vision is supported by three strategic objectives: *Food for People,* to create and accelerate sustainable increases in the productivity and production of healthy food by and for the poor; *Environment for People,* to conserve, enhance, and sustainably use natural resources and biodiversity to improve the livelihoods of the poor in response to climate change and other factors; and *Policies for People,* to promote policy and institutional change that will stimulate agricultural growth and equity to benefit the poor, especially rural women and other disadvantaged groups.

The collaborative work of the CGIAR over the past 40 years has resulted in development impacts on a scale that is without parallel in the international community. They are the result of "international public goods," including improved crop varieties, better farming methods, incisive policy analysis, and associated new knowledge. These products are made freely available to national partners, who transform them into locally relevant products that respond effectively to the needs of rural households in developing countries.

In response to a rapidly changing global development environment, the CGIAR had gone through a major reform to further improve its delivery of research results and on-the-ground impact. The reform is designed to give rise to a more results-oriented research agenda, clearer accountability across the CGIAR, streamlined governance, and increased efficiency. A new CGIAR fund will improve the quality and quantity of funding by harmonizing donor contributions, while a consortium structure will unite the CGIAR Research Centers under a legal entity that provides the fund a single entry point for contracting centers and other partners to conduct research under results-based performance agreements. The shift to a more programmatic approach provides for centers to operate within a strategy and results framework, aimed at strengthening collaboration for greater efficiency and development impact. A portfolio of CGIAR research programs was developed that remains the centerpiece of the reform.

Total contributions to the CGIAR in 2011 were approximately $706 million, with $384 million channeled through the fund. Of that, about 80 percent was untied aid, evidence of widespread faith in a multilateral approach to funding agricultural research for development. This is only a year after the new CGIAR fund was established.

Source: CGIAR Fund Council.

TABLE 5.6   Multilateral development bank progress on Paris Declaration survey indicators
Percentage

Paris Declaration survey indicators (development partner performance)	Global target (%)	Overall results 2010 (%)	MDB performance 2010 (%)	Progress since 2005?
3. Aid flows are aligned with national priorities (aid on budget)[a]	85	41	59	=
4. Strengthen capacity by coordinated support (technical assistance)	50	57	71	+
5a. Use of country PFM systems	55	48	70	+
5b. Use of country procurement systems	50[b]	44	48	+
6. Strengthen capacity by avoiding parallel project implementation units	−67	−32	−73	+
7. Aid is more predictable[a]	71	43	51	−
8. Aid is untied	89	86	100	=
9. Use of common arrangements or procedures (program based approach)	66	45	56	+
10a. Joint missions to the field	40	19	27	+
10b. Joint country analytic work	66	43	56	=

Sources: World Bank and OECD-DAC.
Note: MDBs include African Development Bank, Asian Development, Inter-American Development Bank, and the World Bank. Data are for all participating countries, except for indicator 7 on aid predictability. +, −, and = respectively denote improved, deteriorated and unchanged performance.
a. Unweighted average.
b. AAA target of at least 50 percent is used.

predictability and transparency of flows.[11] The forward-spending plans can be used to provide a good indication of actual CPA disbursements and provide a rough indication of the prospects for total bilateral ODA flows in the coming years.

Country programmable aid is a core subset of ODA (representing about 60 percent of total DAC gross bilateral ODA) and is considered critical support in achieving the MDGs. CPA is aid that has a direct development impact and upon which recipient countries have, or could have, some input—and for which donors are expected to be accountable for delivering.[12] Planned disbursements are reported for the upcoming three years (with latest available forward plans ending in 2013, based on surveys conducted from December 2010 through February 2011). Forward CPA is intended to reduce uncertainty about aid flows in recipient countries and thus enable better management of government spending plans, improve recipient country ownership, and help reveal gaps in development aid.

As a predictor of actual disbursements, planned CPA has proven reliable. For example, the predictability ratio of 2010 flows that were programmed in early 2008, early 2009, and early 2010 relative to actual CPA 2010 disbursements averaged 95.3 percent (OECD 2011c).

Latest available OECD DAC forward survey data indicate that the annual average growth rate of CPA may decelerate in real terms from 4.9 percent during 2001–10 to 2.1 percent during 2011–2013[13] (albeit this represents a recovery from the 0.7 percent contraction in 2010) (figure 5.10). Planned disbursements by multilateral agencies account for much of the 2 percent increase expected during the coming years. In comparison, annual bilateral CPA from the DAC countries is expected to grow by a more modest 1.3 percent. Additionally, actual disbursement rates (versus planned) for CPA declined in 2010, suggesting that the actual pace of growth may be weaker than the anticipated 2 percent rate (OECD 2011).

In per capita terms, CPA is projected to decline by an annual 0.2 percent. Countries in conflict or fragile situations are on track to see a sharper decline in CPA disbursements of 2.1 percent a year on a per capita basis—although they are expected to continue to

FIGURE 5.10   Country programmable aid

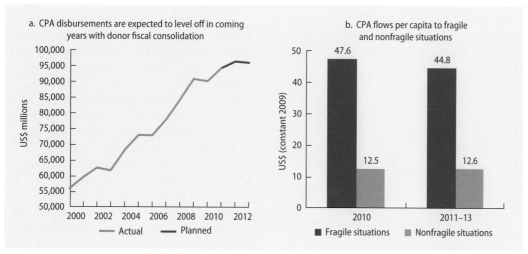

Source: OECD CPA.

receive about four times the per capita CPA disbursements expected for nonfragile countries—underscoring calls at the Busan High Level Forum to scale up aid for fragile and conflict-affected countries and offset this expected decline.

Lower CPA flows could have significant fiscal implications for the countries affected—particularly for those that rely heavily on ODA for external financing needs—and potentially on the achievement of the MDGs. Aid flows are much more significant as a source of external financing for low-income countries. As noted, on average during 2005–10, ODA represented more than 60 percent of total external financing for low-income countries in contrast to a mere 4 percent for middle-income countries, where private financing accounted for more than 60 percent of external financing needs.

During 2011–13, the share of each developing region in total CPA is expected to remain broadly stable. The biggest shifts are expected for Latin America and the Caribbean, where the share is on track to decline from 9.6 percent of the total in 2010 to 8.7 percent on average, and in South Asia, which is expected to see a 1.1 percentage point increase to 21.3 percent on average (table 5.7, figure 5.11). More specifically, the latest CPA

plans for the developing regions indicate the following projected trends.

South Asia is expected to post the strongest gains in CPA inflows over the 2011–13 period, with average annual real growth of 7.7 percent. Three of the projected top four country aid recipients across regions during 2011–13 are in South Asia: Bangladesh, India, and Pakistan (the fourth is Vietnam). Projected CPA represents a 4.7 percent annual increase to South Asia and largely reflects strong growth in flows to Bangladesh, India, and Pakistan, more than offsetting declines in planned flows to Afghanistan and, to a lesser extent, Sri Lanka.

East Asia and the Pacific and **Sub-Saharan Africa** are expected to see an average annual real increase in CPA disbursements of 2.2

TABLE 5.7   CPA by region
Percent share of total

Region	2005	2010	2011–13
East Asia and Pacific	18.8	16.8	16.9
Europe and Central Asia	6.7	8.0	7.8
Latin America and the Caribbean	9.6	9.6	8.7
Middle East and North Africa	20.7	10.7	10.6
South Asia	16.8	20.2	21.3
Sub-Saharan Africa	27.4	34.7	34.8
Memo:			
Fragile situations	—	25.3	25.6

Sources: OECD, CPA and World Bank staff calculations.
— = Not available.

FIGURE 5.11   **CPA flows to developing regions**

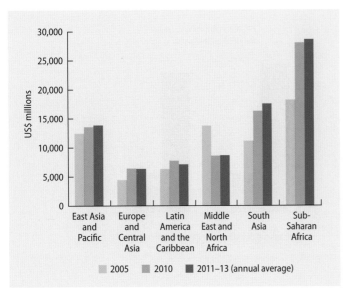

Source: OECD CPA.

percent and 2.1 percent, respectively. For Sub-Saharan Africa, this increase represents a sharp deceleration from the 13 percent annual average increase from 2008 through 2010. On a per capita basis, East Asia and the Pacific is expected to see more modest growth of 1 percent, while Sub-Saharan Africa, given the rapid population growth rate, will see a 3 percent annual decline. Kenya, Ethiopia, Madagascar, and the Democratic Republic of Congo are among the expected top 10 CPA recipients from 2010–13. Vietnam will receive the largest CPA disbursements in East Asia and the Pacific, if plans are realized. Although Indonesia and the Philippines are expected to continue to account for a large share of the CPA flows to East Asia and Pacific, the share is expected to contract compared with 2010.

Latin America and the Caribbean is on track to post the largest real regional decline in CPA through 2013 of close to 7.9 percent a year on average—with the vast majority of the countries in the region recording a contraction. Aid disbursements are expected to decline in nearly 80 percent of partner countries, with 40 percent of these linked to phase-out decisions. On a per capita basis, that translates into a 9.8 percent average

annual decline in planned disbursements for the region.

The **Middle East and North Africa** region is projected to see a modest 0.9 percent expansion of CPA disbursements; however, strong population growth implies an annual average 2.4 percent contraction (in real terms). Planned disbursements to Algeria, Iraq, Jordan, and Tunisia are expected to decline, while those to Egypt and the Republic of Yemen are expected to increase.

Planned CPA disbursements to **Europe and Central Asia** are on track to decline by 0.8 percent overall and by 1.3 percent in per capita terms (at constant 2009 prices). Turkey is expected to continue to post the largest CPA, and modest growth in the share, over the 2011–13 time horizon, while Uzbekistan is expected to see the strongest growth. Bosnia and Herzegovina, Georgia, Kosovo, Moldova, and Tajikistan are among the countries expected to see significant declines.

A significant share of the projected declines in CPA going forward reflect planned phasing-out of aid by donor countries tied to efforts to concentrate aid on fewer partner countries and increased pressures on donor country coffers (OECD 2011c). To reduce transaction costs for recipient countries, where the capacity to manage the administration costs of projects is limited, donor countries have been phasing out programs where disbursements are small. For example, preliminary findings suggest that 162 aid relations between DAC EU member and partner countries are expected to be phased out between 2011 and 2013, accounting for 8 percent of DAC EU total CPA in 2009. From a partner country perspective, while a given donor may not provide large aid volumes in terms of total aid received, it might nevertheless represent a sizable share of aid directed to a specific sector or region where only a few donors may be present.

Real CPA flows to low-income countries are set to decelerate markedly, from an average rate of expansion of 8.6 percent during 2008–10 to 1.4 percent over 2011–13. Planned CPA disbursements to middle-income countries would shift from an

FIGURE 5.12   **CPA received by number of MDG targets achieved or on track**

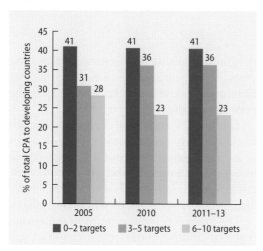

Sources: OECD CPA and DAC, and World Bank staff calculations.

FIGURE 5.13   **CPA by low- and middle-income countries, 2003–13**

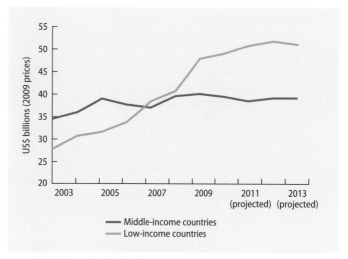

Sources: OECD CPA and World Bank staff calculations.

average real growth rate of 2.2 percent in the earlier period to a 0.2 percent average rate of contraction during 2011–13 (figure 5.12). CPA flows to both low- and middle-income countries are expected to contract on a per capita basis. The largest share of CPA flows are expected to continue to be directed to the countries that are furthest from attaining the MDGs (figure 5.13).

The projected decline in the growth of CPA disbursements likely reflects the need for significant fiscal consolidation in many high-income countries. Among the 23 OECD donor countries, 9 have fiscal deficits equivalent to or greater than 5 percent of their GDP.[14] These countries accounted for 57 percent of bilateral OECD disbursements in 2010 and contributed 22 percentage points to the 63 percent real increase in aid flows from 2000 to 2010 (or $28 billion of the $49 billion level increase). There is also some indication of a slight decline in public support for development assistance. For example, the share of respondents in a post-crisis Eurobarometer survey that considered development important or fairly important fell from 91 percent in 2004 to 88 percent in 2009 (figure 5.14). Nevertheless, according to the same survey, when asked about honoring or going

FIGURE 5.14   **Eurobarometer surveys**

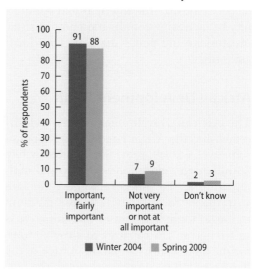

Source: Eurobarometer 2009.
Note: Survey respondents were asked, "In your opinion, is it very important, fairly important, not very important, or not at all important to help people in developing countries?"

beyond existing aid commitments to the developing world, support for development cooperation remained strong: 72 percent of Europeans were in favor of honoring or going beyond existing aid commitments, while only 7 percent deemed that current contribution levels were "too high."

Annex   IFI responses to food price spikes

International financial institutions (IFIs) have responded to recent food price hikes through different lending and nonlending mechanisms (table A5.1).

These responses include emergency financial support to the most vulnerable countries; medium-term assistance to strengthen social safety nets and agribusiness; and long-term programs to enhance infrastructure, rural development, and productivity along the food value chain.

Several high-level meetings in 2008 and after, in addition to the already established Committee on Food Security and the recently created United Nations High Level Task Force, of which the World Bank is a member, helped galvanize the international community by increasing coordination among sister institutions and policy dialogue with local authorities.

African Development Bank

The African Development Bank (AfDB) established the Africa Food Crisis Response initiative in 2008, providing approximately $3 billion to reduce food poverty and malnutrition in the short term and to ensure sustainable food security in the medium to longer term. The aim of this initiative is to strengthen the capacity to closely monitor the food security situation in each of the bank's member countries through the collection, analysis, and dissemination of food security information; to boost sensitization among the different stakeholders in member countries on the dangers, but also on the potential opportunities, that high food prices entail; and to provide budgetary support to low-income food-deficient countries experiencing large fiscal and current account deficits to strengthen food safety-nets for the most vulnerable. In the medium to long term, the objective is to help member countries design and implement national food security programs to ensure production of major food crops, while supporting alternative income-generating activities in rural areas for the poorest segments of the population.

The AfDB's support is channeled through its Agriculture Sector Strategy, which seeks to increase agricultural productivity, enhance incomes, and improve food security on a sustainable basis. It does this through the implementation of two mutually reinforcing pillars.

The first pillar focuses on rural infrastructure—including water resources management and storage, agroprocessing, and trade-related capacities for accessing local and regional markets—as a means of increasing agricultural productivity and food security. In line with the AfDB's principles of strategic focus and selectivity, 80 percent of the 2011 total approvals for the sector were allocated to rural infrastructure.

The second pillar aims to improve the resilience of the natural resource base. Its focus is threefold, namely. forestry, sustainable land management, and climate change mitigation and adaptation. Accordingly, the AfDB recently approved a $63 million grant to support agricultural research on four crops (cassava, maize, rice, and wheat) that African heads of state defined as strategic for the region, through the Comprehensive African Agricultural Development Program.

Asian Development Bank

The Asian Development Bank (ADB) has sought mainly to address the structural and long-term problems associated with food insecurity. The ADB's medium-term investments on food security aim to ease the structural constraints pertaining to productivity, connectivity, and resilience of its developing member countries' food systems. As part of its response, the ADB continued to provide financing for critical agriculture and food security research programs by international

TABLE 5A.1   Responses from the international donor community to recent food price spikes

Institution	Emergency support	Long-term programs
African Development Bank	*Africa Food Crisis Response Initiative:* $730 million for increased provision of agricultural inputs through emergency budget support, use of high-yield New Rice for Africa (NERICA), allocation of resources to fragile states (2008).	*Africa Food Crisis Response Initiative:* $2.2 billion for agricultural infrastructure, including water mobilization for irrigation, rural access roads, and facilities for reducing post-harvest losses (2008).
Asian Development Bank	$700 million for food safety net measures, emergency food assistance, and food policy reforms in Bangladesh, Cambodia, Mongolia, and Pakistan (2007–08).	*Operational Plan for Sustainable Food Security in Asia and the Pacific:* $6.8 billion lending and nonlending assistance allocated to transport and communications, agriculture and natural resources, natural resources management, and rural infrastructure (2009–11).
European Bank for Reconstruction and Development		*EBRD Agribusiness Strategy:* investments in the private sector along the food value chain to foster productivity growth, enhance global food security, and limit food price inflation. $1.3 billion provided (debt and equity) to private agriculture enterprises (2011).
Inter-American Development Bank		$1.8 billion approved for agriculture and rural development over the period 2009–11 (including $551 million for food and agriculture in 2011). $26 million technical assistance projects on concessional terms for small and vulnerable countries in 2011. The IDB Food Security Strategic Thematic Fund ($3.5 million) to provide assistance to Bank borrowing member countries to improve agricultural production and productivity as a means to enhance their food security (supply side).
World Bank	*Global Food Crisis Response Program:* $2 billion (extended through June 2012) to provide financial assistance, policy, and technical advice to the poorest and most vulnerable countries (2008–12). *Global Food Initiative (IFC):* $600 million in investment lending and $300 million in advisory services to support agribusiness value chain in IDA and IDA/IBRD (blend) countries.	Scale up of regular lending program in agriculture and social safety nets. Commitments to agriculture in 2011 reached $3.6 billion. Commitments to social safety nets accounted for $2.9 billion. *Global Agriculture and Food Security Program:* $20 billion financing mechanism to manage the G-20's increased support to agriculture and food security. The program is implemented as a Financial Intermediary Fund for which the Bank serves as trustee (launched in April 2010).

Sources: World Bank and partner institutions.

and national agricultural research centers. At the same time, it actively carried out strategic studies to inform and influence relevant policy making, and to promote regional collective actions for sustainable food security. In carrying out these financing and advisory services, the ADB worked effectively with the Food and Agriculture Organization of the United Nations, the International Fund for Agricultural Development, the International Food Policy Research Institute, the International Rice Research Institute, and other partners.

The ADB's assistance to meet food security concerns became more strategic and focused during the food crisis of 2007–08. The continued uptrend in food prices and forecast of more frequent food price surges prompted it to develop its Operational Plan for Sustainable Food Security in Asia and the Pacific in 2009. From 2009 to 2011 the ADB has provided food-security-related lending and nonlending assistance of $6.8 billion. About $3 billion was allocated to transport and communications, mainly roads, followed by agriculture and natural resources ($2.0 billion) comprising mainly irrigation, drainage, and flood control, water-based natural resources management, and agriculture and rural sector development. By region, ADB food security investments were $2.7 billion to South Asia, $1.8 billion to East Asia, $1.1 billion to Southeast Asia, $888 million to Central and West Asia, and $146 million to Pacific countries.

From January to December 2011 ADB's food-security-related lending and nonlending assistance amounted to $2 billion, $1.8 billion of which went to agriculture and natural resources, energy, transport, and communication. The remaining $200 million was invested in education, finance, industry and trade, public-sector management, and multi-sector activities. Of these investments, $739 million was allocated to South Asia, $470 million to Central and West Asia, $310 million to South East Asia, $307 million to East Asia, and $105 million to Pacific countries.

European Bank for Reconstruction and Development

As the single largest investor in agriculture in its countries of operations, the European Bank for Reconstruction and Development (EBRD) takes action through both debt and equity investments, and complements this investment with technical cooperation and policy dialogue.

In agribusiness, EBRD has adopted a food value chain approach, mobilizing investment from farming and processing to logistics and retail, entirely through the private sector. Ongoing projects include direct support to the primary agriculture sector and leading companies with strong links to the sector. Particularly relevant in response to recent food price surges are the EBRD's activities in improving farmers' risk management and enabling access to seasonal finance through

warehouse receipt and crop receipt programs. The value chain approach recognizes the instrumental role that the supply side in general, and infrastructure and trade logistics in particular, can play in smoothing price variations on international commodity markets.

Even amid the food and financial crises, the EBRD signed 59 projects in 2009, committing €639 million across central Europe and Central Asia. Of this, 42 percent was committed to crisis response projects, with emphasis on supporting low-income and early transition countries.

In 2010 the EBRD scaled up its investments and completed 63 transactions for a record €836 million. In 2011 it surpassed the previous year's volume, providing private agriculture enterprises with debt and equity in the amount of €945 million in transactions.

In addition, EBRD offers a range of instruments that help manage the financial risk of exogenous shocks such as those associated with commodity price volatility. These instruments include policy loans with contingent credit lines, catastrophe risk financing instruments, and interest rate hedges using stand-alone swaps. In parallel with investment, these products allow tailored risk management along the food value chain.

EBRD has been actively engaged through the Private Sector for Food Security Initiative in inducing regulatory and institutional changes in six main areas: promoting public-private sector policy dialogue to achieve greater policy transparency and coordination through the establishment of regular public-private working groups (for example the Ukrainian Grain Sector Working Group); promoting collateralization of soft commodities through technical assistance (implementation of warehouse receipt legislation in Russia, Ukraine, Kazakhstan, and Serbia); improving commodity trading and risk management; enhancing quality standards along the whole food value chain through private and public-sector engagement; increasing local currency financing options; and piloting water audits and policy advise on water-efficient production technology.

Inter-American Development Bank

The Inter-American Development Bank (IDB) established the Food Security Fund in 2008 to address the consequences of major price hikes of food that erupted in 2007. The fund was originally conceived as a two-tiered response to the food-price crisis: in the short to medium term, it helped alleviate the impact of the crisis on the most vulnerable people of the region; in the long term, it aimed to increase agricultural and agro-industrial output and address trade-related policy issues. The fund has recently been refocused on the longer-term objective of improving agricultural production, productivity, and trade as a means to enhance food security. The new Food Security Strategic Thematic Fund will provide technical assistance on concessional terms. Since 2009, 18 technical assistance projects have been approved for a total of $11.9 million. Eleven of these operations were approved for small and vulnerable countries.

Food security has been made a priority area for the IDB as part of the mandates of its Ninth Capital Replenishment (IDB-9). One of the five priority areas of IDB-9 is "Protecting the environment, responding to climate change, promoting renewable energy, and enhancing food security." The IDB-9 Results Framework established a specific target in this regard: by 2015, 5 million farmers should have access to improved agricultural services and investments. Nearly 1 million farmers were assisted in 2010; 2.5 million were assisted in 2011.

Moving forward, the IDB will continue to concentrate on productive activities in order to improve the supply response in the longer term. The IDB's investments toward agriculture growth in the region have quintupled in the past five years, from an average of less than $100 million a year between 2004 and 2006 to nearly $500 million a year for 2009–2011. The IDB's strategic focus is twofold: to increase access to improved agricultural services and rural infrastructure, and to enhance the quality and efficiency of agricultural direct payments. Climate change aspects are considered across all activities directed toward agriculture.

In parallel, the IDB will work to develop new instruments that address the negative impact of price volatility on food security. As part of the G-20's focus on agricultural price volatility, the IDB has joined with the World Bank, Agence Française de Développement, and the International Fund for Agriculture and Development to collaborate in an exchange of information on successful policies and instruments.

World Bank

Responding to the severity of the 2008 crisis and the need for prompt action, the World Bank set up the Global Food Crisis Response Program (GFRP) in May 2008 to provide Bank financing and technical advice to affected countries. The GFRP has now reached 40 million people in 47 countries.

Investment in agriculture and rural development remains a high priority. The World Bank Group is boosting agriculture and agriculture-related investment to some $6 billion to $8 billion a year from $4.1 billion in 2008. In April 2010, at the request of the G-20, the World Bank launched the Global Agriculture and Food Security Program—a multilateral mechanism in support of agriculture that takes up where emergency and recovery assistance leaves off, targeting transformative and lasting change in the agriculture and food security of poor countries through financial support to existing aid effectiveness processes. To date, seven countries and the Gates Foundation have pledged about $1.1 billion over the next three years, with $612 million received.

The World Bank has responded to the food crisis around five main areas:

Policy advice. The Bank has engaged in policy dialogue with more than 40 countries, at their request, to help them address the food crisis. Instruments used include rapid country diagnostics, high-level dialogue, public communications, and in-depth analytical work. A study on sources of food price inflation and appropriate policy responses in Ethiopia is ongoing. In the Middle East and North Africa region, the World Bank, in collaboration with the Food and Agriculture Organization and the International Fund for Agriculture Development, released a paper on "Improving Food Security in Arab Countries."

Expedited financial support. In May 2008, the Bank's Board of Executive Directors endorsed the GFRP, initially a $1.2 billion rapid financing facility providing financial assistance as well as policy and technical advice to the poorest and most vulnerable countries. The Bank increased the size of the facility to $2 billion in April 2009, and the program was recently extended until June 2012 to allow for a swift response to calls for assistance from countries hard hit by price spikes. As of January 2012, the GFRP had financed operations amounting to $1.5 billion; some 82 percent of funds had been disbursed, reaching at least 40 million vulnerable people in 47 countries. In addition to Bank resources, grant funding has been made available through three externally funded trust funds that amounted to about $358 million. A Multi-Donor Trust Fund has received contributions from Australia ($A50 million), Spain (€80 million), the Republic of Korea (W9.5 billion), Canada (Can $30 million), and the International Finance Corporation (IFC) ($150,000). The Russia Food Price Crisis Rapid Response Trust Fund has allocated $15 million for the Kyrgyz Republic and Tajikistan. Last, the European Union has allocated €111.8 million to operations in 10 countries.

Increased IFC investment in agribusiness. The Action Plan projects an increase in support from the World Bank Group (IDA, IBRD, Special Financing, and IFC) to agriculture and related sectors to between $6.2 and $8.3 billion annually over FY10–12. For FY11, IFC invested $2.1 billion across the agribusiness value chain. This leads to a total of $5.7 billion for overall World Bank Group lending in FY11.

Financial market insurance products and risk management strategies. In developing countries, farmers, agro-enterprises, and governments can employ a range of technical, managerial, and financial approaches to mitigate, transfer, and cope with risks. The World Bank supports the development and implementation of agricultural sector and supply chain risk management strategies in a growing number of developing countries through the provision of technical assistance, capacity transfer, and training.

Research to address critical knowledge gaps. In collaboration with other agencies and institutions, the Bank is undertaking a comprehensive analytical program. In addition, the Bank continues its support to the Consultative Group on International Agricultural Research (CGIAR). A new CGIAR Multi-Donor Trust Fund was established to harmonize donor investments and is being hosted and managed by the World Bank. Six new results-oriented research programs submitted by the Consortium of International Agricultural Research Centers have been recently approved for funding by the CGIAR Fund Council.

The World Bank is also responding to the food crisis in coordination with development partners. The World Bank is actively engaged with the United Nations High Level Task Force on the Global Food Security Crisis. Established in April 2008, the task force brings together the heads of UN specialized agencies, funds, and programs with the Bretton Woods institutions. The World Bank is providing financial support to the task force secretariat through the World Bank's Development Grant Facility and also participated in the updating of the UN's Comprehensive Framework for Action. The World Bank is also contributing to several agricultural and food security working groups drafting recommendations for the G-20, at the request of the French presidency. Several G-20 initiatives to address food price volatility are being implemented in collaboration with partners,

BOX 5A.1   Food price hikes and nutrition: The United Kingdom's response

The 2008 food price hike prompted the international community and partner governments to take a number of steps to develop a more coordinated and comprehensive response to undernourishment. As the risks to improved nutrition from high food prices and continued volatility became more apparent, so too did the concern about the lack of progress in tackling hunger and undernourishment (MDG 1).

Along with other governments, the United Kingdom, led by its Department for International Development (DFID), developed a strategic approach to undernourishment based on an evidence paper ("Nutrition and Development: The Evidence"), which culminated in a position paper ("Scaling Up Nutrition: The UK's Position Paper on Nutrition") published in September 2012. These reflect the international policy consensus that undernourishment is best addressed through efforts that reach children in their first 1,000 days of life before the effects are irreversible, and that a twin-track approach is needed which scales up nutrition-specific interventions, often delivered by the health sector, in combination with nutrition-sensitive investments in agriculture, social protection, gender empowerment, and water and sanitation, specifically designed to improve nutrition.

The United Kingdom actively supports the Scaling Up Nutrition movement, which brings together international partners including civil society and the private sector, with country governments to accelerate progress in reducing undernourishment. DFID's

target for 2015 is to reach 20 million children under the age of 5 years with nutrition-related interventions. The agency is scaling up a range of programs across sectors as well as making significant investments in research and impact evaluation to address some of the key evidence gaps. For example:

- In Nigeria, a new six-year program aimed at significantly increasing the coverage of nutrition-specific interventions (treatment of severe acute malnutrition, support to infant and young child feeding, and micronutrient supplementation) in the northern states.
- In Zambia, DFID is providing 10 years of support to the government's child grant program aimed at addressing the economic barriers to good nutrition for children under five.
- In Bangladesh, DFID will strengthen the nutritional impact of existing extreme poverty programs by integrating the delivery of nutrition specific interventions to enhance the impact of the asset transfers, cash transfers, training and income generation which these programs already provide.
- A key research priority is to develop a better understanding of the relationship between nutrition outcomes and agricultural growth, including the nutritional impacts of investments in food staples versus other food crops.

Source: DFID.

including the Agricultural Market Information System launched to improve global agricultural market transparency. The World Bank also regularly participates in the Multilateral Development Banks' Working Group on Food and Water Security.

Bilateral agencies have also undertaken major initiatives to respond to the food crisis, as well. Such efforts include, the United Kingdom's Department for International Development (box A5.1) and the European Union (box A5.2).

BOX 5A.2   EU initiatives on agriculture, food security, and nutrition

The European Commission remains a committed partner both politically and financially to ensuring global food security and nutrition. In terms of financial initiatives for food security, EU cooperation with developing countries is mainly delivered through country programs such as the European Development Fund, where support to agriculture, rural development, and food security is over €1 billion for Africa alone (2008–13). This support is often complemented by other means such as the €1 billion EU Food Facility (2009–11) and the almost €1.7 billion Food Security Thematic Programme (2007–13).

The Food Facility is a prime example of the European Union's ability to react rapidly, efficiently, and transparently to a global food security crisis. This temporary instrument was created as a rapid and specific response to help millions of people in the worst affected countries in the short and medium term, following the food price crisis of 2007–08.

In late 2010, to speed up progress on the MDGs, the European Union announced a €1 billion initiative to assist those countries struggling to reach the MDG targets; this effort focuses on the MDGs that are most off track, in particular MDG 1—eradication of hunger and malnutrition.

As the world's largest grant donor, the European Union is living up to the pledge made in 2009 in L'Aquila to support agriculture and food security with $3.8 billion in 2010–12; in 2010 alone, the European Commission had already committed over 50 percent ($2.02 billion) of its pledge.

The European Commission is taking concerted action on food security. It is a major contributor to global food security governance—especially through its backing for reform of the Committee on World Food Security and for implementation of the Food Security agenda in the G8/G20 context.

Food security is also a priority topic in the EU-US Development Dialogue, with interesting initiatives taking place on the ground. The European Commission has also signed onto a new Strategic Framework of Cooperation, encouraging greater collaboration between several international food agencies and stressing the need for them to focus on their areas of expertise.

The launch of a global initiative to tackle undernutrition and boost efforts to achieve MDG 1.c on malnourishment culminated in the Scaling Up Nutrition (SUN) initiative following months of technical work by experts worldwide, including representatives from the Commission and EU member states (see chapter 2).

Source: European Union.

Notes

1. The remaining disbursements are for "others," made up of more advanced developing countries and territories and amounts unspecified by country.

2. Countries that reached the goal are Ireland (0.52 percent of GNI), Finland (0.55 percent), United Kingdom (0.57 percent), Belgium (0.64 percent), Netherlands (0.81 percent), Denmark (0.91 percent), Sweden (0.97 percent), and Luxembourg (1.05 percent). Others raised disbursements but did not reach the target: France (0.50 percent), Spain (0.43 percent), Germany (0.39 percent), Portugal (0.29 percent). Greek (0.17 percent) disbursements remained constant as a share of GNI, while disbursements from Austria and Italy declined as a share of GNI from 0.52 percent and 0.29 percent in 2005 to 0.32 percent and 0.15 percent in 2010, respectively.

3. The GNI for the 35 low-income countries (of a total of 139 developing countries with CPA data) represented a mere 2 percent of total developing-country GNI in 2010.

4. IDA defines "fragile situations" as either IDA-eligible countries with a harmonized average Country Policy and Institutional Assessment (CPIA, made by World Bank staff each year) country rating of 3.2 or less (or no CPIA), or the presence of a United Nations or regional peacekeeping or peace-building mission, or both, during the past three years. On average, countries in fragile situations reported a CPIA

of 2.9 in 2010, compared with 3.6 by non-fragile-situation countries.

5. For example, Kharas et al. (2011) estimate that net ODA delivered to fragile states rose to an average of $50.4 per capita during 2005–08 from an average of $21.4 during 1995–98, per capita disbursements to non-fragile states remained stable at $10 per capita.

6. The driving force behind these forums has been the Working Party on Aid Effectiveness, hosted at the OECD Development Assistance Committee. What started as a donor-only grouping in 2003 has now emerged as a major international partnership, a forum where donors, developing countries, international organizations, civil society organizations, parliaments, and the private sector meet to discuss how to improve ways to work together.

7. The number of Paris Declaration indicators is 12, but 3 indicators have subindicators, bringing the total to 15. However, 2 indicators do not have global targets, so there are 13 indicators with global targets.

8. Estimates of the associated costs of the unpredictability of aid range between 10 percent and 20 percent of developing-country programmable aid from the European Union over recent years (Kharas 2008).

9. "A New Deal for Engagement in Fragile States." 2011. http://www.oecd.org/data oecd/35/50/49151944.pdf.

10. "Busan Global Partnership for Effective Development Cooperation." 2011. www.aid effectiveness.org/busanhlf4/images/stories/ hlf4/OUTCOME_DOCUMENT_-_FINAL_ EN.pdf.

11. The CPA survey was intended as a means to gain a better understanding of donors' aid allocation policies and to track aid commitments made by the G-8 at Gleneagles.

12. More specifically, CPA is calculated by starting with gross ODA flows, and then excluding aid that is inherently unpredictable (such as humanitarian aid and debt relief); entails no flows to the recipient country (such as donor administrative costs and donor costs of development awareness and research); and is usually not under discussion between the donor agency and partner governments (such

as food aid, aid from local governments, aid through secondary agencies). Additionally, CPA does not net out loan repayments, because these are not typically factored into aid allocation decisions (OECD 2010).

13. The 2011 CPA survey includes forward aid plans for all DAC countries and the largest 23 multilateral agencies, including multilateral development banks, UN agencies, and global funds.

14. The countries are Canada, Greece, Ireland, Japan, the Netherlands, Portugal, Spain, the United Kingdom, and the United States.

References

Adugna, Abebe, et al. 2011. "Trends and Opportunities in a Changing Landscape." CFP Working Paper 8, World Bank, Concessional Finance and Global Partnerships, Washington, DC (November).

Atisophon, Vararat, Jesus Bueren, Gregory De Paepe, Christopher Garroway, and Jean-Philippe Stijns. 2011. "Revisiting MDG Cost Estimates from a Domestic Resource Mobilization Perspective." Working Paper 306, OECD, Paris (December).

Birdsall, Nancy. 2010. "Wow: Will This Results-Based Approach Change DfID Country Allocations?" Center for Global Development, Washington, DC (December).

Birdsall, Nancy, and Homi Kharas. 2010. "Quality of Official Development Assistance Assessment." Brookings Institution and the Center for Global Development, Washington, DC.

de Hennin, Carlo, and Harm Rozema. 2011. "Study on Results-Based Programming and Financing in Support of Shaping the Multi-Annual Financial Framework after 2013." European Union (March).

Eurobarometer. 2009. "Development Aid in Times of Economic Turmoil," Special Eurobarometer for the European Commission, http:// ec.europa.eu/public_opinion/archives/ebs/ ebs_318_en.pdf.

Fengler, Wolfgang, and Homi Kharas. 2010. *Delivering Aid Differently—Lessons from the Field*. Washington, DC: Brookings Press.

Gates, Bill. 2011. "Innovation with Impact: Financing 21st Century Development."

Speech delivered at the Cannes G-20 Summit (November).

Hudson Institute. 2010, 2011. "The Index of Global Philanthropy and Remittances."

Kharas, Homi. 2008. "Measuring the Cost of Aid Volatility." WP3, Brookings Institution, Washington, DC.

———. 2010. "The Hidden Aid Story: Ambition Breeds Success." Brookings Institution, Washington, DC.

Kharas, Homi, et al 2011. "Overview: An Agenda for the Busan High-Level Forum on Aid Effectiveness."

Mwase, Nkunde, and Yongzheng Yang. 2012. "BRICs' Philosophies for Development Financing and Their Implications for LICs." WP/12/72, International Monetary Fund, Washington, DC.

OECD (Organisation for Economic Co-operation and Development). 2009a. "Aid Orphans: Whose Responsibility?" Development Brief, Issue 1, Paris.

———. 2009b. "PARIS21 at Ten: Improvements in Statistical Capacity since 1999" (October). http://www.paris21.org/sites/default/files/P21-at-10.pdf.

———. 2010. "Getting Closer to the Core—Measuring Country Programmable Aid." Development Brief, Issue 1, Paris.

———. 2011a. "Aid Effectiveness 2005–2010: Progress in Implementing the Paris Declaration." Paris.

———. 2011b. "Conflict and Fragility: International Engagement in Fragile States, Can't We Do Better?" Paris.

———. 2011c. "2011 OECD Report on Aid Predictability: Survey on Donors' Forward Spending Plans 2011–2013. Draft." Paris (November).

———. 2011d. "2011 OECD Report on Division of Labour: Addressing Cross-Country Fragmentation of Aid." Paris.

World Bank. 2011. *World Development Report 2011: Conflict, Security and Development.* Washington, DC.

Zimmermann, Felix, and Kimberly Smith. 2011. "More Actors, More Money, More Ideas for International Development Co-operation." *Journal of International Development* 23: 722–38.

Zoellick, Robert. 2011. "Beyond Aid." Speech delivered at George Washington University, Washington, DC (September).

Appendix   Classification of Economies by Region and Income, Fiscal 2012

East Asia and Pacific		**Latin America and the Caribbean**		**South Asia**		**High-income OECD economies**
American Samoa	UMC	Antigua and Barbuda	UMC	Afghanistan	LIC	Australia
Cambodia	LIC	Argentina	UMC	Bangladesh	LIC	Austria
China	UMC	Belize	LMC	Bhutan	LMC	Belgium
Fiji	LMC	Bolivia	LMC	India	LMC	Canada
Indonesia	LMC	Brazil	UMC	Maldives	UMC	Czech Republic
Kiribati	LMC	Chile	UMC	Nepal	LIC	Denmark
Korea, Dem. Rep.	LIC	Colombia	UMC	Pakistan	LMC	Estonia
Lao PDR	LMC	Costa Rica	UMC	Sri Lanka	LMC	Finland
Malaysia	UMC	Cuba	UMC			France
Marshall Islands	LMC	Dominica	UMC	**Sub-Saharan Africa**		Germany
Micronesia, Fed. Sts.	LMC	Dominican Republic	UMC	Angola	LMC	Greece
Mongolia	LMC	Ecuador	UMC	Benin	LIC	Hungary
Myanmar	LIC	El Salvador	LMC	Botswana	UMC	Iceland
Palau	UMC	Grenada	UMC	Burkina Faso	LIC	Ireland
Papua New Guinea	LMC	Guatemala	LMC	Burundi	LIC	Israel
Philippines	LMC	Guyana	LMC	Cameroon	LMC	Italy
Samoa	LMC	Haiti	LIC	Cape Verde	LMC	Japan
Solomon Islands	LMC	Honduras	LMC	Central African Republic	LIC	Korea, Rep.
Thailand	UMC	Jamaica	UMC	Chad	LIC	Luxembourg
Timor-Leste	LMC	Mexico	UMC	Comoros	LIC	Netherlands
Tonga	LMC	Nicaragua	LMC	Congo, Dem. Rep.	LIC	New Zealand
Tuvalu	LMC	Panama	UMC	Congo, Rep.	LMC	Norway
Vanuatu	LMC	Paraguay	LMC	Côte d'Ivoire	LMC	Poland
Vietnam	LMC	Peru	UMC	Eritrea	LIC	Portugal
		St. Kitts and Nevis	UMC	Ethiopia	LIC	Slovak Republic
Europe and Central Asia		St. Lucia	UMC	Gabon	UMC	Slovenia
Albania	UMC	St. Vincent and the		Gambia, The	LIC	Spain
Armenia	LMC	Grenadines	UMC	Ghana	LMC	Sweden
Azerbaijan	UMC	Suriname	UMC	Guinea	LIC	Switzerland
Belarus	UMC	Uruguay	UMC	Guinea-Bissau	LIC	United Kingdom
Bosnia and Herzegovina	UMC	Venezuela, RB	UMC	Kenya	LIC	United States
Bulgaria	UMC			Lesotho	LMC	
Georgia	LMC	**Middle East and North Africa**		Liberia	LIC	**Other high-income economies**
Kazakhstan	UMC	Algeria	UMC	Madagascar	LIC	Andorra
Kosovo	LMC	Djibouti	LMC	Malawi	LIC	Aruba
Kyrgyz Republic	LIC	Egypt, Arab Rep.	LMC	Mali	LIC	Bahamas, The
Latvia	UMC	Iran, Islamic Rep.	UMC	Mauritania	LMC	Bahrain
Lithuania	UMC	Iraq	LMC	Mauritius	UMC	Barbados
Macedonia, FYR	UMC	Jordan	UMC	Mayotte	UMC	Bermuda
Moldova	LMC	Lebanon	UMC	Mozambique	LIC	Brunei Darussalam
Montenegro	UMC	Libya	UMC	Namibia	UMC	Cayman Islands
Romania	UMC	Morocco	LMC	Niger	LIC	Channel Islands
Russian Federation	UMC	Syrian Arab Republic	LMC	Nigeria	LMC	Croatia
Serbia	UMC	Tunisia	UMC	Rwanda	LIC	Curaçao
Tajikistan	LIC	West Bank and Gaza	LMC	São Tomé and Principe	LMC	Cyprus
Turkey	UMC	Yemen, Rep.	LMC	Senegal	LMC	Equatorial Guinea
Turkmenistan	LMC			Seychelles	UMC	Faeroe Islands
Ukraine	LMC			Sierra Leone	LIC	French Polynesia
Uzbekistan	LMC			Somalia	LIC	Gibraltar
				South Africa	UMC	Greenland
				Sudan	LMC	Guam
				Swaziland	LMC	Hong Kong SAR, China
				Tanzania	LIC	Isle of Man
				Togo	LIC	Kuwait
				Uganda	LIC	Liechtenstein
				Zambia	LMC	Macao SAR, China
				Zimbabwe	LIC	Malta
						Monaco
						New Caledonia
						Northern Mariana Islands
						Oman
						Puerto Rico
						Qatar
						San Marino
						Saudi Arabia
						Singapore
						Sint Maarten (Dutch part)
						St. Martin (French part)
						Taiwan, China
						Trinidad and Tobago
						Turks and Caicos Islands
						United Arab Emirates
						Virgin Islands (U.S.)

Source: World Bank data.
Note: This table classifies all World Bank member economies, and all other economies with populations of more than 30,000. Economies are divided among income groups according to 2010 GNI per capita, calculated using the World Bank Atlas method. The groups are: low income, $1,005 or less; lower middle income, $1,006–$3,975; upper middle income, $3,976–$12,275; and high income, $12,276 or more.

ECO-AUDIT
Environmental Benefits Statement

The World Bank is committed to preserving endangered forests and natural resources. The Office of the Publisher has chosen to print *Global Monitoring Report 2012: Food Prices, Nutrition, and the Millennium Development Goals* on recycled paper with 30 percent post-consumer waste, in accordance with the recommended standards for paper usage set by the Green Press Initiative, a nonprofit program supporting publishers in using fiber that is not sourced from endangered forests. For more information, visit www.greenpressinitiative.org.

Saved:
- 21 trees
- 8 million BTU of total energy
- 2,111 pounds of CO_2 equivalent of greenhouse gases
- 9,517 gallons of wastewater
- 603 pounds of solid waste